Contents

Foreword		**3**
Element 1	**The foundations of health and safety leadership**	**6**
1.1	**Reasons for health and safety leadership, organisational health and safety vision and benefits of excellent health and safety leadership**	**7**
	What is health and safety leadership?	7
	The reasons for, and benefits of, effective health and safety leadership	8
	Behaviours/traits of a good health and safety leader	10
	Developing an agreed health and safety vision for an organisation (health and safety leadership value 1)	10
	Building and promoting a shared health and safety vision (health and safety leadership value 1)	11
	The characteristics that make a good health and safety leader	12
1.2	**The moral, legal and business reasons for good health and safety leadership**	**13**
	Moral	13
	Legal	14
	Content for UK students	15
	Content for international students	26
	Business	27
1.3	**Leadership team assurance that health and safety is being managed effectively**	**29**
	Context of the organisation	29
	Risk profiling	29
	Management system thinking	30
	Leadership team involved, informed and visible	33
	Governance, competency and resource	33
	Approval and monitoring of performance indices	33
	Horizon scanning	34
	Benchmarking of organisational health and safety performance	34
1.4	**The influence of good health and safety leadership on health and safety culture**	**35**
	The meaning of safety culture	35
	Promoting fairness and trust in relationships with others (health and safety leadership value 4)	37
	Environmental, health and safety (EHS) and management as a conduit for change	38
	Blame culture, no name no blame and just culture	38
	Three-aspect approach to health and safety culture	40
	Levels of maturity in health and safety culture	41
	Leading and lagging indicators of health and safety culture	43
	Measuring the right things	44
	High Reliability Organisations (HROs)	45

Element 2	Human failure and decision making	**48**
2.1	**The influence of human failure on health and safety culture**	**49**
	Errors	50
	Violations	51
	Impact on safety culture	52
	Providing support and recognition (health and safety leadership value 3)	53
2.2	**Decision making processes, mental shortcuts, biases and habits**	**54**
	The differences between 'Automatic' and 'Reflective' decision making	54
	Reliable mental shortcuts	56
	Individual risk perception	58
	Common biases and how they affect decision making.	60
	Habits and decision making	64
	Personal beliefs and how these can affect decision making	65

Element 3	Leadership	**67**
3.1	**Leadership styles**	**68**
	Transformational leadership	68
	Transactional leadership	70
	Authentic leadership	71
	Resonant leadership	72
3.2	**The five leadership values and supporting foundations**	**75**
	Involvement and communication	76
	Effective role modelling	76
	Embedding	76
	Being considerate and responsive (health and safety leadership value 2)	77
	Assessing own health and safety leadership performance	78
3.3	**Building relationships with the workforce**	**79**
	Leadership walkabouts and rapport	79
	Barriers to building a good rapport with the workforce	83
	What good communication looks like	84
	How information can be given	88
	How to gather information	90
	Encouraging improvement, innovation and learning (health and safety leadership value 5)	92
	Positive reinforcement, negative reinforcement and punishment	93

Further reading   **96**

Health and Safety Leadership Excellence

A course book for the NEBOSH HSE Certificate
in Health and Safety Leadership Excellence

Edition 2

References to legislation in this course book are as at 31 December 2023. For the most up-to-date versions of legislation, please see legislation.gov.uk.

Contains public sector information published by the Health and Safety Executive and licensed under the Open Government Licence

Every effort has been made to trace copyright material and obtain permission to reproduce it. If there are any errors or omissions, NEBOSH would welcome notification so that corrections may be incorporated in future reprints or editions of this course book.

© NEBOSH/HSE 2024

All rights reserved.

No part of this publication may be reproduced, stored in a retrieval system, or transmitted in any form, or by any means, electronic, electrostatic, mechanical photocopied or otherwise, without the express permission in writing from NEBOSH.

Foreword

Health and safety is a key performance measure within successful and forward-thinking organisations. Effective leaders understand that health and safety is not just a moral imperative, but also contributes to the achievement of objectives across the organisational spectrum covering finance, operations, compliance and governance.

Productivity improvements, competitive advantage, talent retention and effective risk management are just a few of the things which flow from strong organisational health and safety performance and culture.

Whether it is finance, marketing, human resources, or health and safety, leaders should always seek to develop their high-level understanding within each component part of their organisation in order to monitor and positively influence overall performance. This book follows the syllabus for the NEBOSH HSE Certificate in Health and Safety Leadership Excellence which is designed to support both leaders and aspiring leaders in gaining core understanding of how their behaviours and responsibilities directly impact on health and safety management.

A guide to the symbols used in this course book

Activity
Carry out an activity to reinforce what you have just learned.

Example
Real or imagined scenarios that give context to points made in the text.

Key Terms
Definitions of key terminology.

Assessment Activity
This symbol indicates that part of the assessment is to be undertaken. This must be done individually and not as part of a group activity. The accredited Learning Partner will advise on the time to be allocated for each part of the assessment.

The HSE's five leadership values

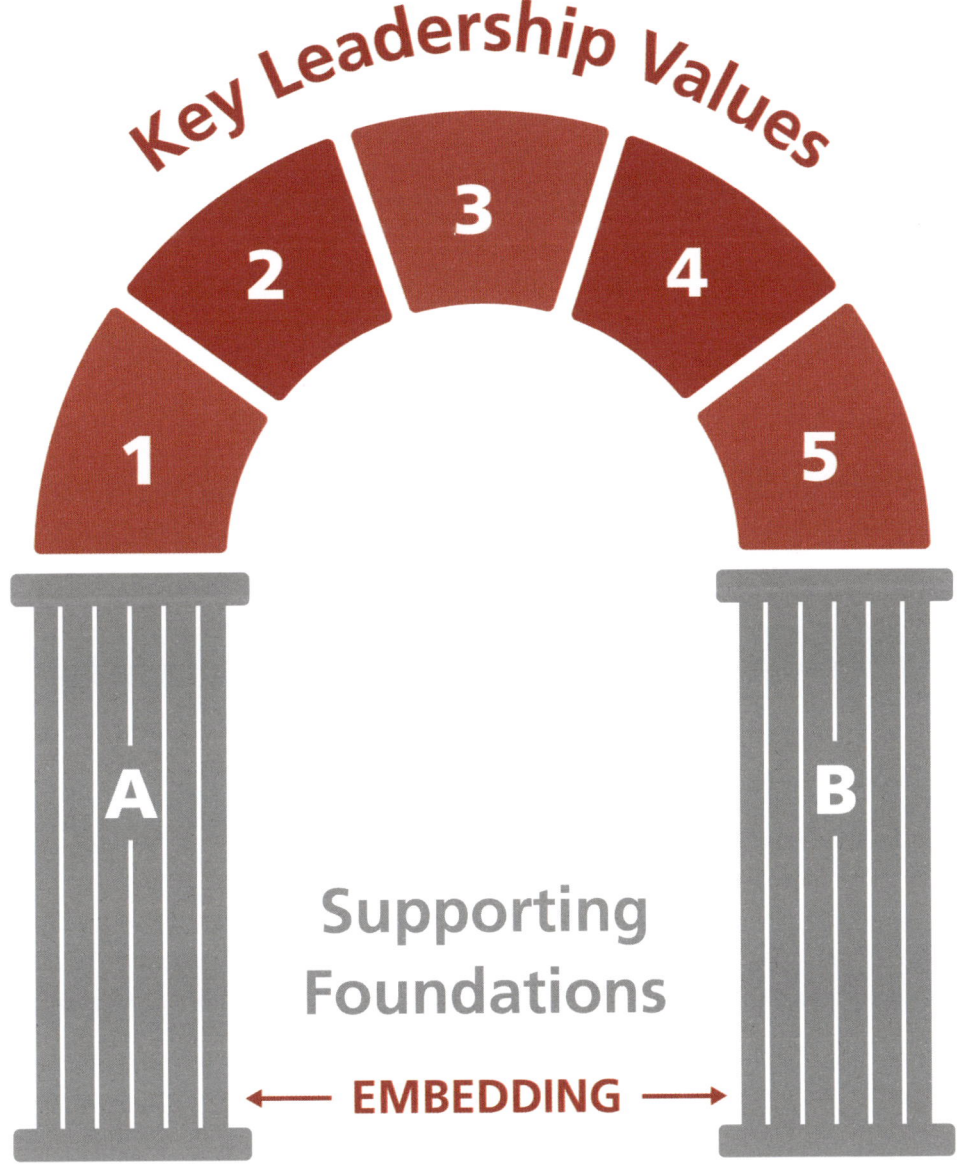

The key leadership values are:
1 Building and promoting a shared H&S vision
2 Being considerate and responsive
3 Providing support and recognition
4 Promoting fairness and trust in relationships with others
5 Encouraging improvement, innovation and learning

The supporting foundations are:
a Involvement / communication
b Effective role modelling

Wherever you see this diagram it indicates that one of the five leadership values will be discussed and assessed.

Element 1
The foundations of health and safety leadership

This Element will explore the reasons for good health and safety leadership, why a health and safety vision is important to an organisation and the benefits that good health and safety leadership can bring to an organisation. To further highlight these areas we will then look at the specific moral, legal and business arguments for good health and safety leadership. Here we will be looking at the level of penalties that organisations and individuals can expect to see should health and safety legislation be breached. The Element will conclude with a look at how leaders can gain assurance that their organisation is managing health and safety well and finally, the impact of good health and safety leadership on organisational health and safety culture.

Learning outcomes
- Demonstrate the importance of health and safety leadership excellence
- Recognise indicators that can provide assurance to leadership teams that health and safety is being managed effectively
- Model good leadership to positively influence health and safety culture

1.1 Reasons for health and safety leadership, organisational health and safety vision and benefits of excellent health and safety leadership

What is health and safety leadership?

Key Term

Leadership can be defined as the process of understanding people's motivations and leveraging them to achieve a common goal.[1]

A debate has long existed about the differences between 'Management' and 'Leadership' across a wider spectrum than health and safety; sometimes it is difficult to determine what these differences might be. However, in the area of health and safety we can draw some clear distinctions about these terms.

It is perfectly possible, and hopefully probable, that a health and safety manager will also be a leader in health and safety. However, this does not mean that a health and safety leader automatically 'manages' the day-to-day functions of organisational health and safety or has ultimate responsibility for them.

When we start to examine the different styles of leadership it will become clear that leadership can come from many places and is not exclusively a 'top down' process. It can be, of course, but leadership is mainly about the ability to 'take people with you' and this skill can be present at all levels. It is, therefore, important that a good health and safety leader will not only have the necessary technical knowledge and skills, but that they also have 'soft skills' such as an open approachable personality, emotional intelligence, and empathy.

Effective health and safety performance comes from the top; members of the board have both collective and individual responsibility for health and safety. Leaders need to examine their own behaviours, both individually and collectively and, where they see that they fall short, to change what they do to become more effective leaders in health and safety.

It is important that leaders take action because:
- protecting the health and safety of workers or members of the public who may be affected by workplace activities is an essential part of risk management and must be led by senior leaders/boards;
- failure to include health and safety as a key business risk in board decisions can have catastrophic results. Many high-profile safety cases over the years have been rooted in failures of leadership; and
- health and safety law places duties on organisations and employers, and directors can be personally liable when these duties are breached: members of the board have both collective and individual responsibility for health and safety.[2]

Activity

When you are working with a good health and safety leader, what makes you think that they are good? How do they exhibit the qualities of a good health and safety leader?

Element 1 The foundations of health and safety leadership

1.1 Reasons for health and safety leadership, organisational health and safety vision and benefits of excellent health and safety leadership

The reasons for, and benefits of, effective health and safety leadership

The reasons for, and benefits of, good health and safety leadership should be self-evident in any organisation; however, it is unfortunately not always the case.

> **The sad but true fact is that many organisations only get to learn the true cost of poor health and safety after an incident has occurred. Many high-profile incidents trace their root causes back to failures of leadership at the very top and many result not just in high penalties, but cause the business to collapse due to irreparable reputation damage.**
> Dame Judith Hackitt, the past Chair of HSE.

If organisations do not regulate themselves, things are likely to go wrong and this is when the regulator will get involved. Investors want to know that they are investing in a well-led/run organisation that performs well. Customers expect organisations to evidence good health and safety performance; there are many companies who will not include organisations in their supply chain who cannot evidence good health and safety management.

An organisation's reputation can be built over many years but it can be destroyed in seconds. Any adverse incidents will have a major impact on all stakeholder interactions which may affect productivity, investment and sales and which may also result in regulatory action..

Even if health and safety law did not exist, there are still sound business reasons for investing in good health and safety. If health and safety is not managed well by its leaders, and things do go wrong, this will have a massive effect on the organisation. As well as the financial (business) reasons, which we will discuss in more detail later, it could mean that the business is unable to continue trading; this will, obviously, have a major impact on workers and other stakeholders.

In the UK, there is a well-established health and safety regulatory system. Companies and individuals can, therefore, face serious consequences when health and safety leadership falls short of what is required. Sanctions include fines, imprisonment and disqualification.

Activity
What do you think the reasons and business benefits are for effective health and safety leadership?

Element 1 The foundations of health and safety leadership

1.1 Reasons for health and safety leadership, organisational health and safety vision and benefits of excellent health and safety leadership

The following case studies from the HSE show what can happen with poor and good leadership.

Examples

The HSE has highlighted a number of examples of weak health and safety leadership. One such case is where a worker at a recycling firm was fatally injured when the machinery that he had been maintaining was not properly isolated and started up unexpectedly.

Commenting on the case, HSE's investigating principal inspector said, "Evidence showed that the director chose not to follow the advice of his health and safety adviser and instead adopted a complacent attitude, allowing the standards in his business to fall".

The HSE and police investigation revealed that there was no safe system of work for maintenance and that instruction, training and supervision were inadequate. As a result, the company director was convicted of manslaughter and received a 12-month custodial sentence.[3]

Conversely, the HSE offers the example of British Sugar as a case study in the benefits to be gained by organisations through robust health and safety leadership.

The HSE says British Sugar had historically had an excellent safety record but in 2003 there were three fatalities at the company. Although health and safety had always been a business priority, the company recognised that a change in focus was needed. This included:
- the CEO assigning health and safety responsibilities to all directors;
- creating effective working partnerships with workers, trade unions and others;
- overseeing a behavioural change programme; and
- annual health and safety targets, and initiatives to meet these.

The results of the leadership-led changes included a two-thirds reduction in both lost times and minor injury frequency rates over a 10-year period, as well as much greater understanding by directors of health and safety risks.[4]

1.1 Reasons for health and safety leadership, organisational health and safety vision and benefits of excellent health and safety leadership

Behaviours/traits of a good health and safety leader

So what qualities should a good health and safety leader evidence? The behaviour of a health and safety leader is as or is more important than their attitude. A good health and safety leader will display behaviour that can influence the workforce. Very often they will not even be aware that they are doing this. Some of these behaviours include:
- encouraging communication (whether this be face-to-face or via other communication methods) with all workers regarding any health and safety anxieties the workers may have;
- making changes to improve working conditions (this is especially useful if the initial idea came from communications with the workforce; it shows that the leader has listened and, more importantly, acted to address the workers' anxieties);
- leading by example by showing the workforce that not only do they know the site health and safety rules but that they also model the correct health and safety behaviours;
- encouraging all levels of the workforce to understand and adhere to the site health and safety rules;
- being actively involved in health and safety committees and taking the lead on any health and safety campaigns/ initiatives that are being introduced;
- where unsafe working practices are taking place, advising the workers concerned about the possible consequences of their unsafe act and that working in such a way is unacceptable.

A failure to exhibit these good behaviours can have a negative effect on the workforce, such as by causing confusion around individual responsibilities caused by unclear or lack of communication.

Developing an agreed health and safety vision for an organisation (health and safety leadership value 1)

Activity
- What would be appropriate for the health and safety vision of your own organisation?
- Does the present version contain or take account of these factors?
- List what you think the most important components might be.

When considering what would be appropriate for the health and safety vision of your organisation, you may have identified items such as:
- a 'speak out' culture where everyone is encouraged to point out issues with health and safety wherever they encounter these;
- a culture where all members of the workforce are inspired to take ownership of health and safety in the organisation; or
- including communication as a vital component in the development of an agreed vision.

Something as vital as an agreed health and safety vision or strategy will not be easy to implement. The effective leader will need to understand and balance points of view, overcome barriers and objections and listen carefully to what all stakeholders have to say, worker input is key.

A vision is about where an organisation wants to get to. A vision is not a strategy; vision is a 'where' and a 'what', while a strategy is a 'how'. In developing the health and safety vision, it may be useful to consider the organisation's values, its stance on health and safety management or on corporate social responsibility. Organisations may also discover that what they want their vision to be is not what it currently is; this is something that leaders in the organisation are empowered to change.

1.1 Reasons for health and safety leadership, organisational health and safety vision and benefits of excellent health and safety leadership

For an organisation's vision to develop and grow, it will ideally build on things that the organisation does successfully. Organisations can be very critical of themselves and sometimes forget to celebrate success. The term 'lessons learned' is not just about stopping things happening again, it is about making sure what is done well is repeated.

The most successful visions tend to be simple, succinct and non-ambiguous. Workers and other stakeholders need to be able to easily understand and support the vision; this is difficult to do if it is overly complex or involved. An overly complex vision is an indication that it is trying to achieve too much and the quality and/or achievements will be diluted as energies and resources will be focused in different directions.

Although a vision should be looking to the future, it should also be time-bound so certain goals can be agreed and ultimately achieved. Each time a goal is achieved that success should be shared and celebrated with stakeholders, they are obviously key in turning the vision into reality.

Building and promoting a shared health and safety vision (health and safety leadership value 1)

The key word in this context is 'shared'. It is vital that everyone feels part of the health and safety vision. It should not be something that people right across the organisation just comply with, it needs to be something to which they contribute. Their contribution must have equal value to all others.

For this to happen effectively will require excellent communication, continuous consultation and discussion, reflection and feedback. Everyone should feel part of the process when establishing goals and objectives for their safety culture.

This should also clearly include shared responsibility; therefore, safety responsibilities need to be clearly defined. This needs to take place across each level of the organisation and should include the policies, goals and plans for the safety culture.

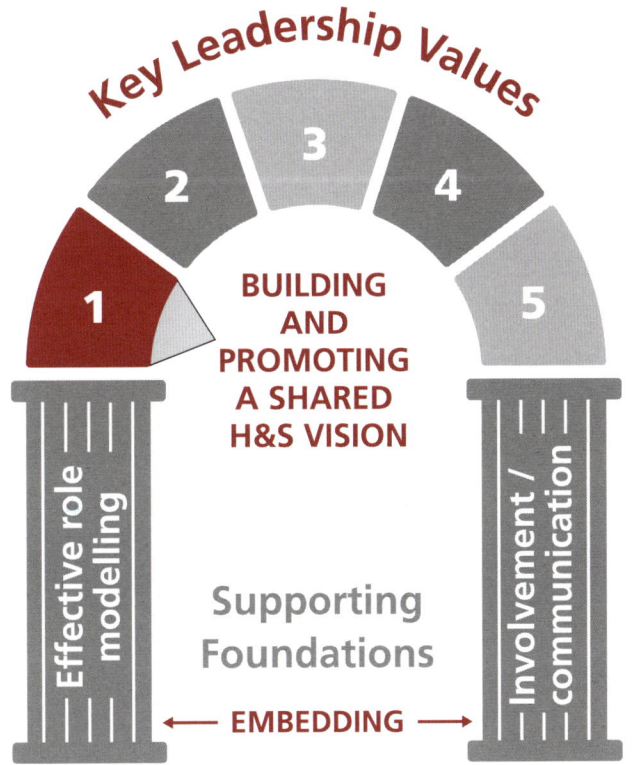

It is vital when building and promoting a shared vision that everyone involved is clear on what is happening, understands why and has the opportunity to comment or challenge when appropriate. It is also important that this health and safety vision is recognised as a long-term process; it should be something worthwhile for people to buy into and not perceived as a 'quick fix' or box ticking exercise for compliance purposes. This will certainly be more credible if the leadership style adopted is responsive and considerate to all those involved.

1.1 Reasons for health and safety leadership, organisational health and safety vision and benefits of excellent health and safety leadership

The characteristics that make a good health and safety leader

This could be a very long list and, unfortunately, can be somewhat discouraging for anyone aspiring to become an effective health and safety leader.

Some of these characteristics include:
- relentlessly driving the health and safety message forward, not as an add-on but as a fundamental business imperative;
- being visible and proactive; and
- being able to articulate important messages across a wide variety of understanding and cultural mix.

Many of these characteristics can be regarded as learned skills and behavious. It is also helpful to understand what leadership really means and this will be discussed later.

Assessment Activity
- Please refer to the document Unit HSL1, guidance and information for candidates and internal assessors

You should now complete task L1: Building and promoting a shared health and safety vision.

Element 1.1 References

1 Chartered Institute of Personnel and Development (CIPD), 'Leadership in the workplace' (https://www.cipd.org/uk/knowledge/factsheets/leadership-factsheet/)
2 HSE, 'Why leadership is important' (https://www.hse.gov.uk/leadership/whyleadership.htm)
3 HSE, 'Case studies: When leadership falls short' (https://www.hse.gov.uk/leadership/casestudies-failures.htm)
4 HSE, 'Case studies: Successful leadership' (https://www.hse.gov.uk/leadership/resources/casestudies.htm)

1.2 The moral, legal and business reasons for good health and safety leadership

Moral

The case for needing good health and safety leadership is often framed in terms of three basic reasons – moral, legal and business.

Societal expectations

There are many social issues that a good leader should be aware of that could affect their organisation.

Government initiatives and campaigns are aimed at the public to raise their understanding of health and safety issues. You may find that health and safety leaders are asked more questions as a result of such initiatives. Some examples of recent campaigns in the UK include the UK Health and Safety Executive's 'Helping Great Britain work well' strategy and campaigns such as the 'Go Home Healthy' campaign.

To try to prevent incidents from occurring in the first place, there is an expectation that health and safety leaders understand their organisation's risk profile. The HSE, in its guidance *Managing for health and safety* (HSG65)[5] states that:

> Effective leaders and line managers know the risks their organisations face, rank them in order of importance and take action to control them. The range of risks goes beyond health and safety risks to include quality, environmental and asset damage, but issues in one area could impact in another.

The risk profile should cover the:
- nature and level of the risks faced by the organisation;
- likelihood of adverse effects occurring and level of disruption;
- costs associated with each type of risk; and
- effectiveness of the controls in place to manage those risks.

Business globalisation and consumer choice also play a role. News stories are easily shared on the internet and social media as well as by traditional media outlets. This can quickly cause damage to an organisation's reputation. Likewise, consumers now have a much bigger say in the products that are on the market with many consumers now preferring brands that have been ethically produced or sourced.

A good health and safety leader should ensure that sufficient control measures are in place to manage the organisation's health and safety risks. At the end of each working day, workers (and their families) understandably expect to return home after being kept healthy and safe while at work.

Responsibility and accountability for health and safety

ISO 45001[6] is an agreed international standard for health and safety management systems. It represents an expectation of best practice. Clause 5.1 requires an organisation's leadership to take "overall responsibility and accountability for the prevention of work-related injury and ill health as well as the provision of safe and healthy workplaces and activities". By taking responsibility for health and safety a good leader sends a clear message to the workforce that they care about the workers' health and safety. As this message flows through the organisation it can lead to positive changes, for example, by improving morale in the workforce which in turn could lead to an improved health and safety culture. Health and safety leaders should aim to make themselves as approachable and visible as possible within the organisation, for example, by conducting regular walkabouts.

Element 1 The foundations of health and safety leadership

1.2 The moral, legal and business reasons for good health and safety leadership

The protection of workers from reprisals when reporting health and safety incidents and hazards

The health and safety management system standard (ISO 45001) contains a clause relating to health and safety leadership. Part of this clause is to ensure the protection of workers when reporting health and safety incidents. As a leader you should be encouraging your workers to report all cases where they believe there is danger from an uncontrolled hazard, or if they have been involved in an incident (accident or near-miss). You should ensure that all managers, team leaders and supervisors understand your organisation's policy on reporting and that they should encourage workers under their control to report incidents. You can do this by, for example, ensuring that systems are in place for reporting (especially a near-miss reporting system), that the system is accessible by all workers, that it is easy and not time consuming to use and by ensuring that this message is cascaded down the workforce. Feedback to the workforce on actions taken regarding reported incidents will go some way to proving to your workforce that you do want them to report and that no reprisals will be taken against them if they do report an incident.

Many countries also have legislation in place to protect their workers when reporting a health and safety incident or a hazard, for example, in the UK workers are protected under section 44 of the Employment Rights Act 1996 (protection from suffering detriment in employment) in health and safety cases.

Legal

The role, function and limitations of legislation as a means of promoting health and safety performance

In simple terms, health and safety legislation is there to help to protect the health and safety of the workforce. It does this by imposing legal duties on employers and workers and a system of penalties (such as fines and imprisonment) for non-compliance. Although it should not be the only driver, the threat of these penalties can be a big motivator for better health and safety performance.

Key Terms

Legislation tends to be either:

goal setting (sets objectives to be met, leaving the detail on exactly how to do this up to the employer); or

prescriptive (tells organisations exactly what to do and when to do it).

Some of the limitations of using legislation to promote health and safety performance include:
- organisations ignoring best practice and only doing enough to meet the legal minimum requirements;
- organisations seeing health and safety as a regulatory and financial burden instead of as a tool to help them protect their workforce;
- legislation (particularly prescriptive legislation) not keeping pace with change and not addressing current issues;
- legislation using language that is unclear and open to interpretation; and
- organisations (especially those with inadequate health and safety assistance) not being familiar with the law.

1.2 The moral, legal and business reasons for good health and safety leadership

Content for UK Students

As a leader there are specific pieces of legislation that can affect you and that you must be aware of. Ignorance is no defence when it comes to the law.

The following legislation will be considered:
- Health and Safety at Work etc Act 1974 (HSWA)
- Company Directors Disqualification Act 1986
- Corporate Manslaughter and Corporate Homicide Act 2007

Health and Safety at Work etc Act 1974

The Act is goal setting. As a leader the main sections of the Act that you need to be aware of are sections 2, 3, 36 and 37. The Act applies in England, Scotland and Wales; Northern Ireland is covered by the Health and Safety at Work (Northern Ireland) Order 1978. Where we refer to section numbers, these refer to the Health and Safety at Work etc Act 1974. The equivalent sections from the Northern Irish Order are as follows:

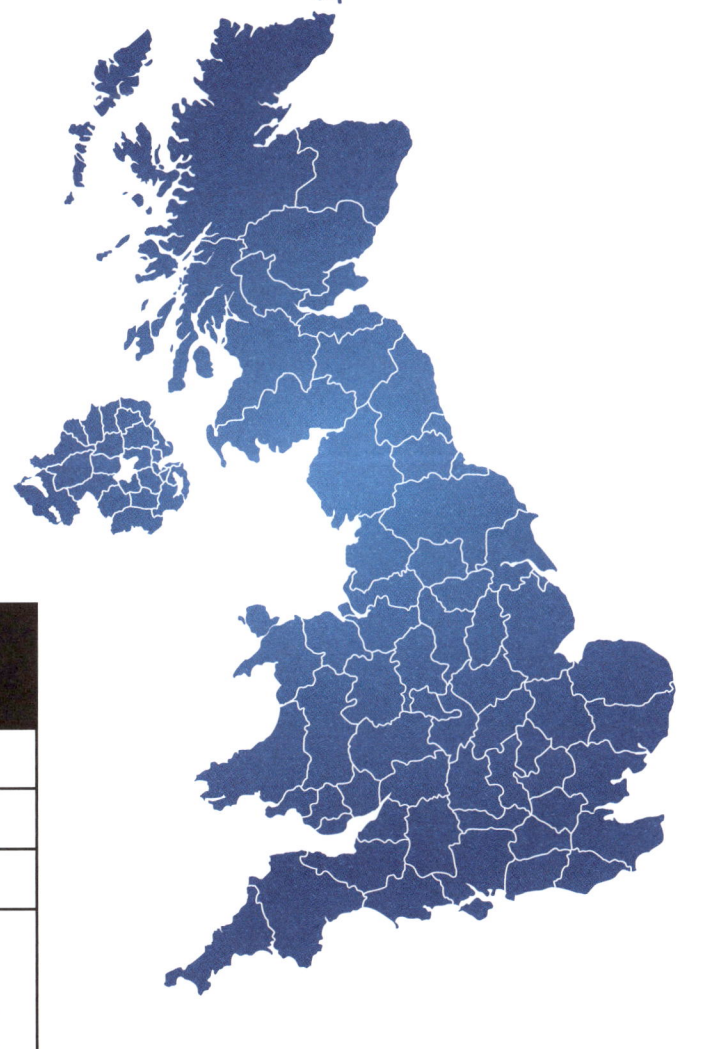

Health and Safety at Work etc Act 1974	Health and Safety at Work (Northern Ireland) Order 1978
Section 2	Section 4
Section 3	Section 5
Section 36	Section 34
Section 37	Covered under section 20(2) of the Interpretation Act (Northern Ireland) 1954. Section 34A of the 1978 Order amends the wording of section 20(2) in relation to health and safety offences.

Key Term
Reasonably practicable
Reasonably practicable requires the employer to balance the costs of compliance with the duty against the risk. This can include time, effort and money. It is the risk that determines whether the cost involved is justified.

Element 1 The foundations of health and safety leadership

1.2 The moral, legal and business reasons for good health and safety leadership

Section 2

The first part of this section puts a duty on employers to protect the health, safety and welfare of all of their workers. The second part of this section imposes specific duties. These are:

1. To provide and maintain safe equipment and systems of work as far as is reasonably practicable.

2. To provide arrangements for ensuring, so far as is reasonably practicable, safe working practices for storage, use, handling and transport of articles and substances.

3. To provide information, instruction, training and supervision to ensure, so far as is reasonably practicable, the health and safety of all workers.

4. To maintain any workplace that is under the employer's control, so far as is reasonably practicable, in a safe condition so that it does not provide risk to the health and safety of all workers. The workplace must also have safe points of access and egress and these points must be maintained.

5. To provide and maintain a working environment that is, so far as is reasonably practicable, safe, without risks to health, and has adequate facilities and arrangements for the workers' welfare at work.

Organisations are also required to have a health and safety policy; if the organisation employs five or more workers this must be a written a policy that states how the organisation is going to manage health and safety. The policy must be brought to the attention of all workers.

Section 2 also requires organisations to consult with workers on health and safety matters. The duty to consult is set out in legislation.

The final part of the section relates to forming a health and safety committee. It is good practice to set up a health and safety committee as this will involve the workforce in organisational health and safety matters.

> **Further information**
> The HSE's website provides further information on health and safety policies,[7] the duty to consult with workers[8] and health and safety committees.[9]

Section 3

Section 3 is very similar to section 2 but it looks at the protection of 'others' who are not directly employed by the organisation but who could be affected by the organisation's work activities. For example, protection of temporary and contract workers, visitors to site and the general public.

The Act also imposes a duty on self-employed people to carry out their work so that other persons are not exposed to risks to their health or safety. However, self-employed people 'whose work activities pose no potential risk of harm to others' are now exempt from this section of the Act.

The Act also requires employers, and qualifying self-employed people, to provide information to relevant parties on the way that they conduct the work activities that might affect the health and safety of others.

1.2 The moral, legal and business reasons for good health and safety leadership

Section 36

This section looks at offences committed by a 'person' (this could be an individual or an organisation) due to the act or fault of some 'other person' (this could be a worker, manager, supervisor, contractor, consultant) then the 'other person' is guilty of the offence. In these cases the HSE could bring a prosecution against the 'person', the 'other person' or both. So, for example, if poor advice from a senior manager to a worker leads to a breach of duty, the senior manager could be prosecuted. The organisation may also be prosecuted but it would depend on the level of culpability. Prosecutions under section 36 are not very common.

Section 37

This section will probably have the biggest impact on leaders; it means that senior directors/managers of an organisation can be individually liable for breaches of health and safety law and can be prosecuted as well as the organisation.

The wording from the Act states that it is for offences made by a 'body corporate' (an organisation). It must be proved that the offence was "committed with the **consent or connivance** of, or have been attributable to any **neglect** on the part of any director, manager, secretary or other similar officers or a person who was purporting to act in any such capacity". **Consent and connivance** is interpreted as having knowledge and making decisions based on that knowledge but turning a blind eye. In these circumstances, **neglect** can include situations where a director ought to have been aware of the circumstances; in other words neglect does not require knowledge.

> ### Example
>
> Recent years have seen a rise in section 37 prosecutions and the sentences handed down. To illustrate this, a section 37 case that was prosecuted in 2023, saw a property development company and a construction company fined.[10] The managing director of the property development company was also given a suspended sentence.
>
> These sentences were given following a construction worker being killed when panes of glass fell on top of them as they were unloading a container. The unloading of the container was not properly planned, supervised or carried out safely. As a result:
> - The managing director of the property development company was found guilty of breaching section 37(1) of the Health and Safety at Work etc Act 1974. They were sentenced to nine months' imprisonment (suspended for two years) and ordered to pay £57,172 in costs.
> - The property development company was found guilty of breaching section 3(1) of the Health and Safety at Work etc Act 1974 and regulation 4(1) of the Work at Height Regulations 2005. They were fined £100,000 and ordered to pay £55,085 in costs.
> - The construction company was found guilty of breaching section 3(1) of the Health and Safety at Work etc Act 1974 and regulation 4(1) of the Work at Height Regulations. They were fined £115,000 and ordered to pay £52,562 in costs.

Element 1 The foundations of health and safety leadership

Enforcement

The HSE can bring prosecutions against the organisation for breaches of sections 2 and 3; these could result in substantial fines for the organisation. As well as prosecutions, the cost of any civil claim brought against the organisation could also be substantial.

The Company Directors Disqualification Act 1986

If a director is found guilty of an offence (in this case section 37 of the Health and Safety at Work Act etc 1974), the court can make an order to disqualify the individual from 'the promotion, formation or management' of another organisation. The maximum period of disqualification is five years for a summary offence or 15 years for an indicatable offence.

Key Terms

England, Northern Ireland and Wales

A **summary offence** is a **less serious** offence and is usually heard in Magistrates' Courts. The maximum sentence that a Magistrates' Court can impose is an unlimited fine and/or up to six months' imprisonment.

An **indictable offence** is a more serious offence and is usually heard in the Crown Court. The maximum sentence that the Crown Court can hand down is an unlimited fine and/or up to two years' imprisonment.

Scotland

In Scotland, offences are triable through two routes – summarily or on indictment. The latter is referred to as 'Solemn'. For offences committed after 16 January 2009, the penalties are set out in the Health and Safety (Offences) Act 2008. The maximum penalties for nearly all offences are, for summary conviction, £20,000 and/or 12 months' imprisonment and, for Solemn, an unlimited fine and/or two years' imprisonment. The offences of obstructing an inspector and impersonating an inspector are summary offences only and the maximum penalties are £5,000 and £5,000 and/or 12 months' imprisonment, respectively.

1.2 The moral, legal and business reasons for good health and safety leadership

Individual duties and possible enforcement actions for gross negligence manslaughter

Gross negligence manslaughter applies to an individual rather than the organisation. The offence is where someone is killed due to another person's extreme recklessness/carelessness. The maximum prison sentence for gross negligence manslaughter is life imprisonment.

There is a four-staged legal test for manslaughter by gross negligence. The following elements must be established:
1 there must be a duty of care owed by the defendant to the deceased person;
2 the defendant must have breached the duty of care;
3 the breach must have caused or significantly contributed to the death of the deceased; and
4 the breach must be characterised as gross negligence and, therefore, considered a crime.

A breach of duty of care happens when an individual, who owes the duty of care, does not act in the same way as a reasonable person would do in the same position. Therefore, if the individual was acting within the range of what was generally accepted as standard behaviour/practice, it will be difficult to prove that they have breached the duty of care.

When looking at work-related deaths, very often proceedings are also taken against the individual under sections 7 (which is not part of this course), 36 or 37 of the Health and Safety at Work etc Act 1974.

Example

In May 2017, a company director was sentenced to 32 months' imprisonment when he admitted causing the death of a golf club worker. The worker was 29 years old and had ADHD and learning difficulties. He died while collecting golf balls from an eight-foot deep lake with a weighted belt and breathing equipment; the breathing equipment was lost during the dive. The golf worker was paid £20 to £40 per day instead of the defendant employing a trained diver that would have cost approximately £1000 day. The director had stood and watched the incident happen and only raised the alarm when he saw a constant stream of bubbles rising to the lake's surface and saw that the floatation device carrying the air supply floated to the side of the lake. The director admitted to manslaughter by gross negligence.[11]

Element 1 The foundations of health and safety leadership

1.2 The moral, legal and business reasons for good health and safety leadership

Sentencing for gross negligence manslaughter cases

Key Term

Culpability refers to how much the defendant is at fault for the offence. This ranges from very high to low. High is where there was a deliberate breach or a flagrant disregard for the law. Low is where it is found that failings were minor and occurred as an isolated incident.

The sentencing guidelines for gross negligence manslaughter were published in 2018 and apply to all cases sentenced on or after 1 November 2018 no matter when the offence occurred. These guidelines apply to England and Wales only. The maximum sentence for gross negligence manslaughter is life imprisonment.

If the accused is convicted, they will be sentenced. When sentencing, the judge will follow the process set out in the sentencing guidelines and will first decide which offence category (culpability range) the offending falls into:

A	B	C	D
Very high	High	Medium	Lower

This decision is based on the offender's culpability and the guidelines set out factors that will indicate each level of culpability. The offence category determines the starting point and category range for the sentence:

	Culpability			
	A	B	C	D
Starting point	12 years	8 years	4 years	2 years
Category Range	10-18years	6-12 years	3-7 years	1-4 years

Where it is not clear which offence category a case falls into, the judge can adjust the starting point. This adjustment is applied before any adjustments for aggravating and mitigating features of the case. Once the starting point has been set, the judge will consider other matters that could increase or decrease the sentence. These include:
- aggravating factors, which may be statutory, such as previous convictions or 'other', such as ignoring previous warnings;
- mitigating factors such as remorse or attempts to assist the victim;
- any assistance given to the prosecution;
- guilty plea(s);
- dangerousness of the offender (when considering a life or extended sentence); and
- if sentencing for more than one offence, the totality principle (is the total sentence proportionate to the overall offending?)

The judge must also:
- consider whether to make any compensation or ancillary orders, such as disqualification as a director;
- give reasons for, and explain the effect of, the sentence; and
- take account of any time spent on bail where tagged curfew applied.

1.2 The moral, legal and business reasons for good health and safety leadership

The Corporate Manslaughter and Corporate Homicide Act 2007

This legislation is about holding the organisation accountable rather than individual directors. The offence of corporate manslaughter (corporate homicide in Scotland) can be brought:
- when the way in which an organisation's activities are managed or organised causes a person's death and amounts to a gross breach of a relevant duty of care owed by the organisation to the deceased; and
- if the way in which its activities are managed or organised by its senior management is a substantial element of breach of duty of care.

The offence is, therefore, aimed at the strategic/top level management of an organisation rather than activities conducted by lower-level workers (in other words the organisation's leaders).

The offence of corporate manslaughter/homicide is an indictable offence which can only be heard in the High Court.

The penalties available to the Court when an organisation is convicted are:
- a fine; and/or
- a remedial order; and/or
- a publicity order.

A remedial order can be made which requires the organisation to remedy:
- the breach of the duty of care;
- anything that looks like it may have caused the death; and
- any deficiencies in the organisation's health and safety systems that the offence highlights.

The Court can also make a publicity order. This requires the organisation to publicise:
- that it has been convicted of the offence;
- the details of the offence;
- the amount of any fine; and
- the terms of the remedial order (if applicable).

The organisation can be ordered to publicise details of the offence on their website or by taking out an advertisement in the local press or in trade publications/websites. If the organisation does not comply with the publicity order this is also an indictable offence and could result in an unlimited fine should the organisation be found guilty.

Examples

The first corporate manslaughter case was sentenced in 2011 following an incident in September 2008. A geologist employed by an organisation was investigating soil conditions in a trench; the trench collapsed and killed the worker. The organisation was successfully prosecuted and received a fine of £385,000 payable over 10 years.[12]

In 2022, an aluminium recycling company pleaded guilty to corporate manslaughter and was fined £2 million (with £105,514 in costs). A worker had sustained fatal head injuries while working near dangerous, unguarded machinery. The machinery had not been isolated before workers conducted cleaning and maintenance work. The company admitted the charge on the basis that their management failings amounted to corporate manslaughter. Were it not for the guilty plea, the fine would have been one-third higher.[13]

1.2 The moral, legal and business reasons for good health and safety leadership

The Health and Safety Offences and Corporate Manslaughter sentencing guidelines

Sentencing in England, Wales and Northern Ireland

The definitive guidelines for sentencing health and safety offences and corporate manslaughter offences are published by the Sentencing Council. They apply to England and Wales only. Courts in Northern Ireland can, however, refer to the guidelines as a starting point when sentencing.

When sentencing for health and safety offences, the first step for the court in sentencing is to identify the initial harm category. This is based on the risk of harm created by the offence. There are four levels, 1 to 4, with one being the most severe. The harm category is based on both the seriousness of the harm risked (levels A to C) and the likelihood (high, medium or low) of that harm arising.

Level A is where a death or injury which will result in lifelong injury has occurred. Level B includes physical or mental impairment, not requiring lifelong care, but which has a substantial and long-term effect on the sufferer's ability to carry out normal day-to-day activities or on their ability to return to work; or a progressive, permanent or irreversible condition. Level C covers all other cases not falling within Level A or Level B.

The following table shows the harm categories based on these levels and the seriousness of harm:[14]

	Level A	Level B	Level C
High likelihood of harm	Harm category 1	Harm category 2	Harm category 3
Medium likelihood of harm	Harm category 2	Harm category 3	Harm category 4
Low likelihood of harm	Harm category 3	Harm category 4	Harm category 4 (start towards bottom of range)

The guidelines give a range of sentences that are appropriate for each type of offence (offence ranges). For each offence there are a number of categories that reflect varying degrees of seriousness. For each category there is a starting point for sentencing.

When deciding on the level of fine the judge must consider the following:
- the level of culpability of the defendant;
- the level of harm created by the offence;
- the number of workers or members of the public exposed to the risk AND whether the breach was a significant cause of actual harm.

1.2 The moral, legal and business reasons for good health and safety leadership

The level of fine is based on the organisation's **turnover**, not profit. There are five categories of organisation:
- micro (turnover of not more than £2million) organisations;
- small (turnover between £2million and £10million);
- medium (turnover between £10million and £50million);
- large (turnover of £50million or more); or
- 'very large organisations' that have a turnover which greatly exceeds £50million can be fined outside of the normal ranges should the offence warrant this.

When sentencing, the judge will look at the starting point for the fine but will also take into account if there are any factors where an increase or reduction in the fine can be made. The following table gives some examples of the level of fine that organisations can receive based on the relevant factors.

Level of culpability	Harm category	Size of organisation	Fine starting point	Maximum fine	Minimum fine
Very high	1	Large	£4million	£10million	£2.6million
High	4	Large	£240,000	£700,000	£120,000
Very high	1	Medium	£1.6million	£4million	£1million
Low	2	Medium	£40,000	£100,000	£14,000
High	3	Small	£54,000	£210,000	£25,000
Medium	2	Small	£54,000	£230,000	£25,000
Very high	1	Micro	£250,000	£450,000	£150,000
Medium	2	Micro	£30,000	£70,000	£14,000

The message from the Definitive Guideline for the sentencing of health and safety offences (the Sentencing Guideline) is clear: the fine must be sufficiently substantial to have a real economic impact which will bring home to both management and shareholders the need to comply with health and safety legislation. The highest fine imposed in 2023, for example, was £4.4 million.

1.2 The moral, legal and business reasons for good health and safety leadership

For corporate manslaughter offences there is an unlimited maximum fine but the offence range is from £180,000 to £20million. An offence that is considered to be level 'A' (where the organisation is deemed to have a high level of culpability) for a 'large organisation' would have a starting point of £7.5million with the category range between £4.8million and £20million.

For health and safety offences, individuals can be prosecuted as well as organisations. As with organisations, fines are based on the culpability of the individual. However, it is also important to note that individuals could also receive a jail term for the most serious offences. Summary offences can attract an unlimited fine and/or up to six months' custodial sentence. An indicatable offence will also attract an unlimited fine but the jail term for these offences could be up to two years.

For example, where an individual is convicted and was found to be very highly culpable, the following sentences (depending on the harm category) could apply.

Harm category	Sentence starting point	Category range
1	18 months' custody	1 - 2 years' custody
2	1 year's custody	26 weeks - 18 months' custody
3	26 weeks' custody	Band F fine* or high level community order - 1 year's custody
4	Band F fine	Band E fine - 26 weeks' custody

* Fines are split into six bands; the starting point for each band is as follows:

- Band A – 50% of relevant weekly income
- Band B – 100% of relevant weekly income
- Band C – 150% of relevant weekly income
- Band D – 250% of relevant weekly income
- Band E – 400% of relevant weekly income
- Band F – 600% of relevant weekly income

Therefore, a director earning £98,000 per annum (£1885 per week) could find themselves with a personal fine starting at around £11,500 when the harm category is 4.

Examples

In January 2017 a retail company was fined £2.2million and ordered to pay costs of £71,000. The fine was due to a worker being injured by a cage falling over which left the worker paralysed below the hip with only a 1% chance of ever walking again. The level of culpability in the case was set at high with a harm category of 2. The organisation's turnover was way in excess of that for large companies so the judge treated this case as a 'very large organisation'. When sentencing the judge took into account the guilty plea and other mitigating circumstances.[15]

Also in January 2017 a food manufacturer was fined £2million and ordered to pay costs of £20,000. A worker sustained a spinal fracture after falling nearly two metres from the top of a mixing machine while attempting to clean it. The company's turnover put them in the 'very large organisation' category but the judge decided to treat the organisation as 'large'. The culpability level was high and the harm category was 1. Based on the definitive guideline, the starting point for the fine would, therefore, have been £2.4million with a category range of between £1.5million and £6million.[16]

1.2 The moral, legal and business reasons for good health and safety leadership

Sentencing in Scotland

Scotland does not currently have any official guidelines but there are 'sentencing factors' which the judge can consider.

The Scottish Sentencing Council have drawn up a wide range of factors which judges should generally consider when deciding on a sentence. The judge will decide:
- which factors presented to the court are relevant and should be taken into account; and
- what weight to give to each factor; these can be aggravating factors (which will make the sentence more severe) and mitigating factors (which will make the sentence less severe).

Some of the general factors which can be considered are:
- the type and seriousness of the crime;
- the culpability of those involved;
- protection of public and deterrence; and
- personal circumstances of the offender.

Aggravating factors that the court will take into account include:
- the effects of the crime on the victim(s); and
- previous convictions.

Where they apply, the factors that the judge will consider to be mitigating include:
- a guilty plea by the offender;
- if this is the offender's first offence; and
- any assistance provided to the prosecution following a guilty plea.

The following table gives an overview of the sentences available in the various levels of Scottish Court (the Procurator Fiscal will decide which court the case will be heard in).

	Justice of the Peace Court	Sheriff Court (summary)	Sherriff Court (solemn)	High Court
Who decides the verdict?	Justice of the Peace	Sheriff	Jury	Jury
Who sets the sentence?	Justice of the Peace	Sheriff	Sheriff	Judge
Maximum fine available	Up to £2,500	Up to £10,000	Unlimited	Unlimited
Maximum length of imprisonment	Up to 60 days	Up to 1 year	Up to 5 years	Up to life

Sheriffs/judges are able to impose a fine and/or imprisonment depending on the offence. The courts can also hand down community-based sentences.

For more information on the sentencing factors please refer to guidance from the Scottish Sentencing Council.[17]

Element 1 The foundations of health and safety leadership

1.2 The moral, legal and business reasons for good health and safety leadership

Content for international students

Different levels of standards and enforcement in different jurisdictions

There are different standards of health and safety around the world; some countries have mature, well-embedded systems while other countries have very rudimentary systems or no system at all. Organisations that operate globally can find this frustrating; what meets the standard in some countries will not in others. Some larger organisations will, therefore, use standards from countries with robust systems to manage their risks wherever they are operating in the world. When tendering for large projects, their systems of work, policies, procedures and so on will reference these higher standards.

These organisations are sometimes seen as leaders within the country of operation; once the local workforce sees that better standards are available they are more likely to start demanding this from other organisations.

For the majority of organisations, the fact that a country has a robust regulatory system in place (including an enforcement regime) will act as a big incentive to provide better health and safety standards for their workforce.

The International Organization for Standardization publishes a standard for occupational health and safety management systems: ISO 45001:2018 *Occupational Health and Safety Management Systems: requirements with guidance for use*. The standard is recognised globally and provides a framework for organisations to use to manage their health and safety risks. As all organisations will have to evidence how they meet each of the clauses within the standard, this can also be seen as a driver to improve health and safety standards globally.

The International Labour Organisation (ILO)[18] has drawn up over 40 conventions and recommendations and produced over 40 codes of practice relating to occupational health and safety. These conventions and codes of practice give the minimum standard that countries and organisations should aim to implement. However, the standards set out within ILO conventions are not always policed within those countries who have ratified the conventions. If there is no policing, this usually means no enforcement actions are taken which, in turn, means that there is no incentive for countries or organisations to improve health and safety standards. There may also be different standards of health and safety around the globe due to countries interpreting the language of the conventions in different ways.

Responsibilities of leaders under Article 20 of the C155 Occupational Health and Safety Convention 1981

The Occupational Health and Safety Convention 1981 requires each member state to have a "coherent national policy on occupational safety, occupational health and working environment". Article 20 of the convention requires "Co-operation between management and workers and/or their representatives within the undertaking…". Health and safety leaders should, therefore, be proactive and be seen to be engaging with the workforce whenever possible. Workers are more likely to co-operate with the employer when leadership is visible and is interested in the workforce.

1.2 The moral, legal and business reasons for good health and safety leadership

Business

The level of fines/penalties/compensation

In the UK, the level of fines imposed on conviction for offences related to health and safety have risen significantly since the introduction of the sentencing guidelines. However, as well as fines, there could also be costs associated with:
- fees charged by regulators;
- putting right anything identified in enforcement notices;
- the cost of non-production should a prohibition notice be served by a regulator;
- loss of business and reputation should any enforcement action be publicised;
- level of compensation/damages due to injured parties.

The real cost of accidents/incidents

Each year millions of working days are lost due to workplace accidents and ill-health. Each year, the UK Health and Safety Executive publishes health and safety statistics.[19] In 2022/2023, there were:
- 1.8 million cases of work-related ill-health
- 135 workers killed in work-related accidents
- 600,000 non-fatal injuries to workers
- 35.2 million working days lost due to work-related illness and workplace injury.

These figures are also reflected globally. The International Labour Organization (ILO) estimates that some 2.3 million people around the world succumb to work-related accidents or diseases every year. This corresponds to more than 6000 deaths every day. Worldwide, there are around 340 million occupational accidents and 160 million victims of work-related illnesses annually.

The annual cost to the global economy is estimated to be 3.94% of global Gross Domestic Product.

There are many costs, other than fines, that should be considered. Some costs can be insured against but the majority of costs incurred are not covered by insurance and must be absorbed by the organisation.

It is estimated that the ratio of insured to uninsured costs is roughly 1:8 (so for every £1 of insurance payment the organisation receives they will pay out a minimum of £8 but could be as much as £36). The analogy that is very often drawn is comparing costs to an iceberg. The tip of the iceberg, visible above water, represents insured costs but the majority of the iceberg, which is hidden under the water, represents the uninsured costs.

Examples of costs that can be insured against are:
- medical costs relating to injury and/or ill-health; and
- damages to the injured party or to the family of a deceased worker.

Element 1 The foundations of health and safety leadership   27

1.2 The moral, legal and business reasons for good health and safety leadership

In the UK the insurable costs are covered through compulsory employers' liability insurance.

Some examples of costs which are uninsurable are:
- delays in production;
- additional wage bills for overtime payments/temporary workers to cover the injured person's job;
- sick pay for the injured person;
- loss of contracts resulting from either loss of reputation and/or being unable to meet orders due to production down-time;
- damage to equipment, plant, products or premises;
- fines and costs orders;
- legal expenses;
- investigation time and site clear up costs; and
- the excess of any insurance claim.

Activity
Despite the moral arguments for high standards of occupational safety and health, why do you think that fines are often seen as a bigger driver for improved performance?

Element 1.2 References

5   HSE, *Managing for health and safety* (HSG65, 3rd edition, 2013) (https://www.hse.gov.uk/pubns/books/hsg65.htm)
6   International Organization for Standarization, 'ISO 45001:2018 *Occupational Health and Safety Management Systems: Requirements with guidance for use*' (2018) (www.iso.org)
7   HSE, 'Prepare a health and safety policy' (https://www.hse.gov.uk/simple-health-safety/policy/index.htm)
8   HSE, 'Consulting and involving your workers' (https://www.hse.gov.uk/involvement/index.htm)
9   HSE, 'Health and safety committees' (https://www.hse.gov.uk/involvement/hscommittees.htm)
10 HSE, 'Director given suspended prison sentence and firms fined after worker dies' (2023) (https://www.shponline.co.uk/corporate-manslaughter/evergreen-construction-limited-leyton-homes-limited-and-a-director-fined-after-worker-dies/)
11 SHP Online, 'Golf company director jailed after lake death' (2017) (www.shponline.co.uk/common-workplace-hazards/golf-company-director-jailed-lake-death)
12 HSP Online, 'First corporate manslaughter conviction delivers £385,000 penalty (2011) (www.shponline.co.uk/legislation-and-guidance/first-corporate-manslaughter-conviction-delivers-385-000-penalty/)
13 *R v Alutrade Limited & Others* (unreported), Wolverhampton Crown Court, 25 March 2022
14 Sentencing Council, 'Health and safety offences, corporate manslaughter and food safety and hygiene offences' (www.sentencingcouncil.org.uk/sentencing-and-the-council/about-sentencing-guidelines/about-published-guidelines/health-and-safety-offences-corporate-manslaughter-and-food-safety-and-hygiene-offences/)
15 SHP Online, 'Wilko fined £2.2M after "horrific workplace accident"' (2017) (https://www.shponline.co.uk/in-court/wilko-fined-2-2m-for-horrific-workplace-accident/)
16 Food manufacturer, 'Warburtons fined £2M for worker's fall' (2017) (https://www.foodmanufacture.co.uk/Article/2017/01/27/Warburtons-fined-for-worker-s-fall)
17 Scottish Sentencing Council, 'Sentencing factors', (www.scottishsentencingcouncil.org.uk/about-sentencing/sentencing-factors)
18 International Labour Organization (https://www.ilo.org/global/lang--en/index.htm)
19 HSE, 'Health and safety statistics' (https://www.hse.gov.uk/statistics/)

1.3 Leadership team assurance that health and safety is being managed effectively

The Health and Safety Executive, in association with the Institute of Directors, produced a set of guidelines for organisations to assist in giving assurance to leadership teams that the organisation's health and safety practices are being managed effectively. These guidelines,[20] provide more detail on this vital component of health and safety culture.

Context of the organisation

Boardroom decisions must be made in the context of the organisation's health and safety policy; it is important to 'design-in' health and safety when implementing decisions. Identifying who is a stakeholder with regards to an organisation is also key; it is not just internal workers who are stakeholders. They can also include suppliers, local communities and customers and consulting with these groups (when appropriate) can provide valuable insights.

Suppliers and contractors also have specialist knowledge that can be vital when introducing change. It is important to remember that often, it is also these people who are the last to know when work-based changes have been introduced, and any changes undertaken can directly affect their activities on a work site.

ISO 45001:2018 has a clause entitled Context of the organisation.

Risk profiling

The risk profile of an organisation should inform all aspects of the approach to leading and managing health and safety risks.

Every organisation will have its own risk profile and effective leaders must know the risks their organisations face, rank them in order of importance and take action to control them. This is the starting point for determining the greatest health and safety issues for an organisation. In some businesses the risks will be tangible and with immediate obvious safety hazards. In other organisations the risks may be health-related and it may be a long time before any illness becomes apparent.

Health and safety leaders need to ensure that their respective organisations have built a risk profile that covers:
- the nature and level of the threats faced by an organisation;
- the likelihood of adverse effects occurring;
- the level of disruption and costs associated with each type of risk; and
- the effectiveness of controls in place to manage those risks.

The outcome of risk profiling will be that the right risks are identified and prioritised for action, controls communicated, and minor risks are not given too much priority. It also informs decisions about the risk control measures that are needed and where resources should be made available and allocated.

Further information on risk profiling can be found in the HSE's publication *Managing for health and safety* (HSG65).[21]

1.3 Leadership team assurance that health and safety is being managed effectively

Management system thinking

Good health and safety management does not happen by accident. Management of health and safety in any type of organisation requires clearly defined processes. An effective health and safety management system will help an organisation meet legal obligations, as it will assist compliance with legislation and any internal corporate standards such as the health and safety management system ISO 45001.

An effective health and safety management system is the product of a structured and focused effort that places health and safety at the centre of business decisions and not as an after-thought. There are many different models of health and safety management system, but all follow the same Plan, Do, Check, Act cycle (PDCA) as part of a continual improvement process. In the PDCA cycle, the following broad steps are taken:
- **plan**: establish a clear set of goals and targets that will move the organisation forward in terms of health and safety management;
- **do**: carry out actions to improve health and safety;
- **check**: monitor and determine whether the steps you have taken are moving you closer to your goals; and
- **act**: take action as a result of the monitoring findings in order to continually improve.

In terms of a health and safety management system (for example, ISO 45001), these steps are broken down under:
- context of the organisation;
- leadership and worker participation (health and safety management system framework);
- planning (plan);
- support (do);
- operation (do);
- performance evaluation (check); and
- improvement (act).

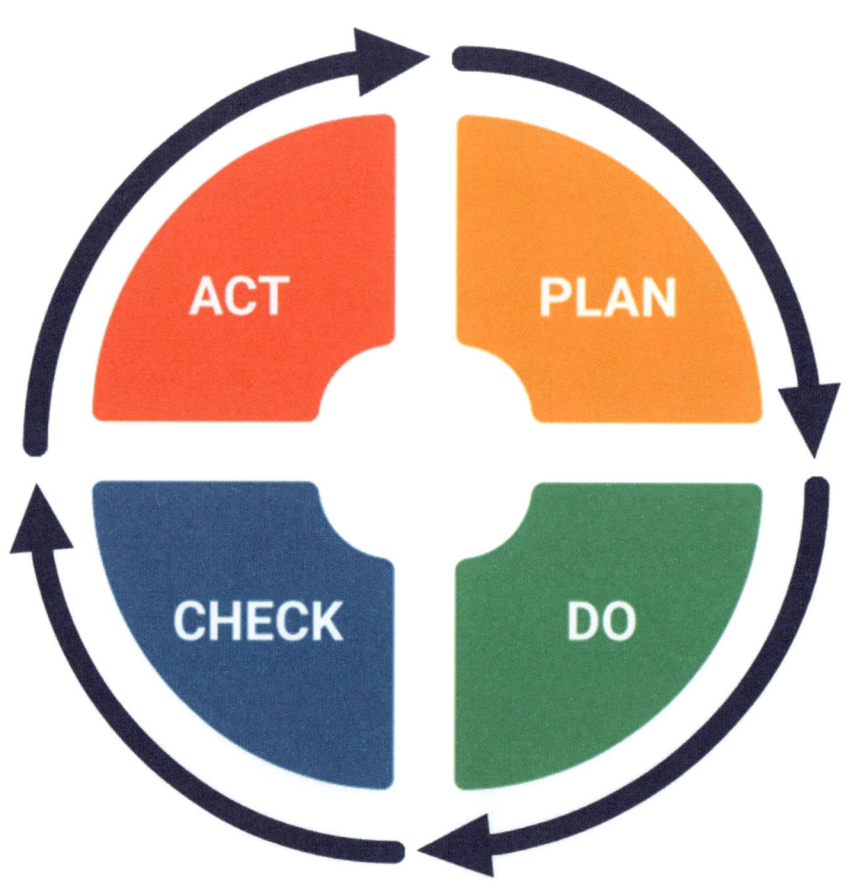

Element 1 The foundations of health and safety leadership

1.3 Leadership team assurance that health and safety is being managed effectively

The sections of a health and safety management system

Leadership (underpinning all clauses of the health and safety management system)

ISO 45001:2018 requires organisations to evidence how their leadership commit to the health and safety management system. There are many points that an organisation must consider that range from taking overall responsibility for health and safety to promoting continual improvement.

If your organisation has a formalised health and safety management system, you may find that, as a health and safety leader, you are invited to take part in both internal and external audits to evidence how you meet this clause.

It is also important that you should be involved in the implementation of the health and safety policy. The policy should be a clear demonstration of the aims of the organisation towards health and safety, together with the vision and commitment to achieve these goals. It should also commit to complying with legislation and to continually improve processes. It should come from board level. As part of the policy you will also be expected to assign roles and responsibilities for health and safety throughout the organisation.

Another important part of the leadership clause of ISO 45001:2018 is consultation with the workforce and gaining worker participation. It is really important that your workers are consulted on anything that could affect their health and safety such as the implementation of new work equipment. You should also engage your workforce and encourage them to take part in health and safety-related activities. For example, get them involved in the risk assessment process; the workers are carrying out their work activities and are, therefore, best placed to help you understand the hazards and risks associated with their work.

Planning (plan)

This should include the processes for identifying hazards and reducing the risks. The plan should then seek to deliver a reduction in risk while prioritising the biggest issues. At the same time, the plan should also identify potential breaches of, or changes in, legislation or compliance obligations, and address these too. You should also ensure that you consider all risks and opportunities associated with the health and safety management system. For example, a risk to the health and safety management system could be an ineffective audit programme; an opportunity could be bringing in technology to improve health and safety performance by automating higher risk activities.

Support (do)

This section of the health and safety management system looks at resources, competence, awareness, communication and document control. It is the health and safety leader's responsibility to ensure that there are adequate resources provided to implement or maintain the health and safety management system. You should also ensure that the workforce is aware of relevant information relating to the health and safety management system, that is that they know what is contained in your organisation's health and safety policy, the implications of not conforming and so on. It is also really important that competence levels of all workers throughout the organisation are determined; if any gaps in knowledge or skills are identified you should ensure that the workers receive appropriate training. Any documents relating to the health and safety management system should be adequately controlled. If your organisation has a formal quality management system (QMS) you can integrate this requirement with the QMS.

Element 1 The foundations of health and safety leadership

1.3 Leadership team assurance that health and safety is being managed effectively

Operation (do)
You will need to ensure that processes (written or physical) are put into place to eliminate hazards with reference to the hierarchy of control (eliminate, substitute, engineering controls, administrative controls, issue of personal protective equipment as a last resort).

The clause also looks at the management of change, whether the change is temporary or permanent. As a leader you should ensure that processes are in place and that any possible consequences of the change have been taken into account.

Organisations are also required to control their contractors from the tendering stage right through to the end of the work. Leaders should ensure that contactors are provided with all relevant information regarding the identified hazards on site. This also refers to outsourced activities.

The final requirement of this clause is emergency preparedness; again, you should be ensuring that your organisation has in place plans to deal with any potential emergencies and that these plans are tested on a regular basis, and updated as and when required.

Performance evaluation (check)
In order to determine if progress is being made towards the goals, there should be monitoring and measurement activities carried out. These will include active and reactive measures (also known as 'leading' and 'lagging' indicators), the most widely used are accident data (lagging) and statistical trends. Leading indicators (such as planned maintenance) are important as they can identify potential problems before an incident occurs. As well as monitoring against the targets, there should also be assessments to determine whether compliance obligations are being complied with.

The organisation should also have in place a robust internal audit programme that should establish the scope and frequency of the audit.

Compliance with legislation and standards has already been discussed. This part of the clause is to ensure that regular compliance evaluations are carried out. So you will need to ensure that you have you identified all relevant compliance obligations AND can evidence how you meet each obligation.

The final part of this clause is for management to carry out regular reviews of the management system to review the data and ensure that the plan remains relevant to feed the continual improvement process. This should lead to action, as necessary, to correct any identified issues.

Improvement (act)
The organisation should be looking to (continually) improve the performance of the health and safety management system. To do this you will need to be looking at any non-conformances that may be identified via audit or some other method such as complaints from workers. You should also ensure that any accidents and (very importantly) near misses are thoroughly investigated. This should help to stop the incident from occurring again in the future. Any additional controls implemented should be reviewed to ensure their continuing suitability.

The organisation should also consider any other improvements that could be made. Suggestions that come from the workforce are usually invaluable; the health and safety leader should not discount these and only expect solutions to come from the top.

1.3 Leadership team assurance that health and safety is being managed effectively

Leadership team involved, informed and visible

Today's workplaces are increasingly complex and the demands on all levels of management, especially senior leaders, grows ever more acute. As a result, leaders need to rely more than ever on the intelligence, resourcefulness and competence of their teams and workers. Collaboration is an essential ingredient within a strong health and safety culture. Collaboration leads to gains in morale, realises creativity, changes attitudes and behaviours and gets buy in to new initiatives and targets. Collaboration also helps to keep leaders informed as to current challenges with regards to health and safety management. More importantly, leaders remain visibly engaged with health and safety initiatives and demonstrate how important they are to organisational objectives.

Being visible and taking the time to be seen is not about checking on workers; effective leaders schedule time to engage with workers, in order to determine that they are receiving the care and attention that they deserve. Witnessing work in action also helps a leader to determine if workers are properly trained. Working side-by-side with workers also gives a leader an opportunity to explore concerns and communicate a health and safety vision in an informal way.

Leadership visibility and involvement can also be demonstrated by:
- health and safety appearing regularly on the agenda for board meetings;
- a chief executive (or other senior leader) visibly demonstrating health and safety leadership;
- having a health and safety director on the board. This gives a strong signal that health and safety is taken seriously and that the board understands its strategic importance; and
- senior leaders setting health and safety targets. This helps define what an organisation is seeking to achieve.

Governance, competency and resource

For many organisations, health and safety is a corporate governance issue. Leaders in these organisations integrate health and safety into the main governance structures, including board sub-committees, such as risk, remuneration and audit. Organisations should have robust systems of internal control, covering not just financial risks but also risks relating to the environment, business reputation and health and safety. Health and safety management must of course be adequately resourced. A famous quote from Trevor Kletz,[22] one of the pioneers of process safety management, is:

> …if you think safety is expensive, try an accident.

This is very true, but more importantly, people also get hurt, and sometimes killed. Damage to workers' health can have a significant impact on their lives.

Resource spent on ensuring competency within the workplace is rarely wasted. A competent person is someone who has sufficient high quality training and experience or knowledge and other qualities that allow them to assist an organisation to achieve its aims. The level of competence required will depend on the complexity of the situation and the particular skill set needed. Certainly, with regards to health and safety management, it is essential competent advice is available.

Approval and monitoring of performance indices

In order to determine how well an organisation is performing with regards to health and safety management, leading and lagging indicators must be measured.

Leaders can use leading and lagging indicators to ensure that:
- appropriate weight is given to reporting both preventive information (such as progress of training and maintenance programmes) and incident data (such as accident and sickness absence rates);
- periodic audits of the effectiveness of management structures and risk controls for health and safety are carried out;

Element 1 The foundations of health and safety leadership

1.3 Leadership team assurance that health and safety is being managed effectively

- the impact of changes such as the introduction of new procedures, work processes or products, or any major health and safety failure, is reported as soon as possible to the board; and
- there are procedures to implement new and changed legal requirements and to consider other external developments and events.

Continuous improvement

All good organisations strive to improve performance. There is evidence that a strong health and safety culture leads to improvements in quality. As already described, following a management review, the last step of any PDCA cycle is continuous improvement by implementing any identified improvements.

Continuous improvement is something that is essential within any organisation intending long-term success. What is essential with regards to health and safety management is how an identified improvement is implemented. By being very clear as to why the change is being introduced, it will be more readily accepted and adopted by workers. Additionally, organisations should have robust procedures in place for any change management process; these should consider the implication of any 'change' prior to it being put into practice.

Horizon scanning

With regards to health and safety management, horizon scanning can be used to detect early signs of potentially important developments and opportunities, with emphasis on new technology and its effects on the workplace. For example, is new equipment being developed that could be introduced to the workplace in order to reduce risk? Organisations also need to be aware of any upcoming changes in law or other compliance obligations, so they can anticipate changes and consider their response.

Benchmarking of organisational health and safety performance

Health and safety benchmarking is used to assess the health and safety performance of an organisation by comparing it with that of other high performing companies within the same industry. Benchmarking allows organisations to identify and implement possible improvements. Organisations that undertake benchmarking achieve the following benefits:

- each organisation learns other ways in which it could improve the delivery of health and safety and ways in which it can plan to implement the improvements;
- each organisation gains a better understanding of their processes for dealing with health and safety that they can build on to develop suitable performance measures and targets; and
- it builds successful partnerships that can aid future health and safety improvements.

Element 1.3 References

20 HSE and Institute of Directors, *Leading health and safety at work* (INDG417(rev1), 2013) (https://www.hse.gov.uk/pubns/indg417.htm)
21 HSE, *Managing for health and safety* (HSG65, 3rd edition, 2013) (https://www.hse.gov.uk/pubns/books/hsg65.htm)
22 See, for example, US Chemical Safety Board, 'News release' (https://www.csb.gov/statement-from-csb-chairperson-rafael-moure-eraso-on-the-passing-of-noted-chemical-process-safety-expert-professor-trevor-kletz/)

1.4 The influence of good health and safety leadership on health and safety culture

Good leadership and effective leadership techniques should have a positive impact on the culture of any organisation. This of course should include the health and safety culture of an organisation. Without positive influential leadership it will prove difficult, or perhaps impossible to ensure the right culture is systematically maintained.

The meaning of safety culture

> **Key Term**
>
> **Safety culture**
>
> The Confederation of British Industry describes the culture of an organisation as "the mix of shared values, attitudes and patterns of behaviour that give the organisation its particular character. Put simply it is 'the way we do things round here'". They suggest that the "safety culture of an organisation could be described as the ideas and beliefs that all members of the organisation share about risk, accidents and ill health".[23]

But what is a safety culture and how do organisations know if they have one?

Safety culture must ultimately be part of a larger organisational culture, but what is that? It has been said that organisational culture is easier to feel or experience than to describe. Unlike organisational structure it does not have a formal recognisable shape. It has sometimes been described as 'the way we do things around here'. Charles Handy has written many books on the subject of culture. His text, *Understanding Organisations*[24] is still used extensively to help try and make sense of how organisational culture works within a structure to influence the way an organisation, as well as all those within it, work together. Clearly the organisational culture will have significant influence over the safety culture.

It is important for health and safety leaders to understand human failures and the differences between errors and violations. (These will be discussed further in Element 2.1.) Human failure is not random; there are distinct patterns and types of human failure with different causes and therefore different ways of addressing them. For example, forgetting an essential step in a task due to fatigue compared with an operator cutting a corner in a procedure due to productivity requirements. As a leader the important thing is to fully understand the root cause of the failure and understand the local context of the individual affected by the failure.

It is important to remember that errors will happen. How you deal with these as a leader could directly influence the safety culture within the organisation. Again, this comes back to a 'just culture' and ensuring that these situations are handled consistently through all levels of the organisation.

The problem with culture, in this case safety culture, is that it is hard to measure, and therefore to understand what needs to be changed and how to go about making the changes so that everyone involved will benefit.

The table that follows provides a checklist of positive actions that underpin a positive safety culture. Broadly speaking this would indicate that an organisation had a good safety culture.

1.4 The influence of good health and safety leadership on health and safety culture

Harm category	this is shown when management…	and is helped when management…
Visible commitment to safety by management	- Make regular useful visits to site - Discuss safety matters with frontline personnel - Will stop production for safety reasons regardless of cost - Spend time and money on safety eg to provide protective equipment, safety training, and conduct safety culture workshops or audits - Will not tolerate violations of procedures and actively try to improve systems so as to discourage violations eg plan work so that short cuts aren't necessary to do the work in time.	- Makes time to visit site (not just following an accident or incident) - All show commitment - Has good non-technical skills eg communication skills; - Are also interested in workforce safety when they are not at work, eg provide information on domestic safety - Shows concern for wider issues eg workforce stress and general health - Actively sets an example eg always conform to all safety procedures

Workforce participation and ownership of safety problems and solutions	- Consults widely about health and safety matters
	- Does more than the minimum to comply with the law on consultation
	- Seeks workforce participation in:
	  ○ setting policies and objectives
	  ○ accident/near miss investigations
		- Supports an active safety committee
		- Have a positive attitude to safety representatives
		- Provides tools or methods that encourage participation eg behavioural observation programmes and incentive schemes that promote safety

Trust between shop floor and management	- Encourages all employees and contractors to challenge anyone working on site about safety without fear of reprisals - Keeps their promises - Treats the workforce with respect	- Promotes job satisfaction/good industrial relations and high morale - Promotes a 'just' culture (assigning blame only where someone was clearly reckless or took a significant risk) - Encourages trust between all employees
Good communications	- Provides good (clear, concise, relevant) written materials (safety bulletins, posters, guidance) - Provides good briefings on current issues day to day and in formal safety meetings; listening and feedback	- Encourages employee participation in suggesting safety topics to be communicated - Provides specific training in communication skills - Has more than one means of communicating
A competent workforce	- Ensures that everyone working on their sites is competent in their job and in safety matters	- Is supportive - Has a good competence assurance system

Source: *HSE Human Factors Briefing Note No 7. Safety Culture* [25]

1.4 The influence of good health and safety leadership on health and safety culture

Promoting fairness and trust in relationships with others (health and safety leadership value 4)

Trust is something that is generally earned. Building a relationship based on trust and fairness requires that a leader not only possesses those qualities but also visibly demonstrates them. There is no point in leaders promoting qualities such as openness, honesty and integrity unless they are prepared to lead using those qualities. Failure to do so opens the way for accusations of hypocrisy and it is difficult to recover from this and built trust once insincerity has been identified. The effective leader in health and safety must lead from the front and by doing, not just by saying. Sharing responsibility for health and safety and delegating is also an important part of being an effective leader.

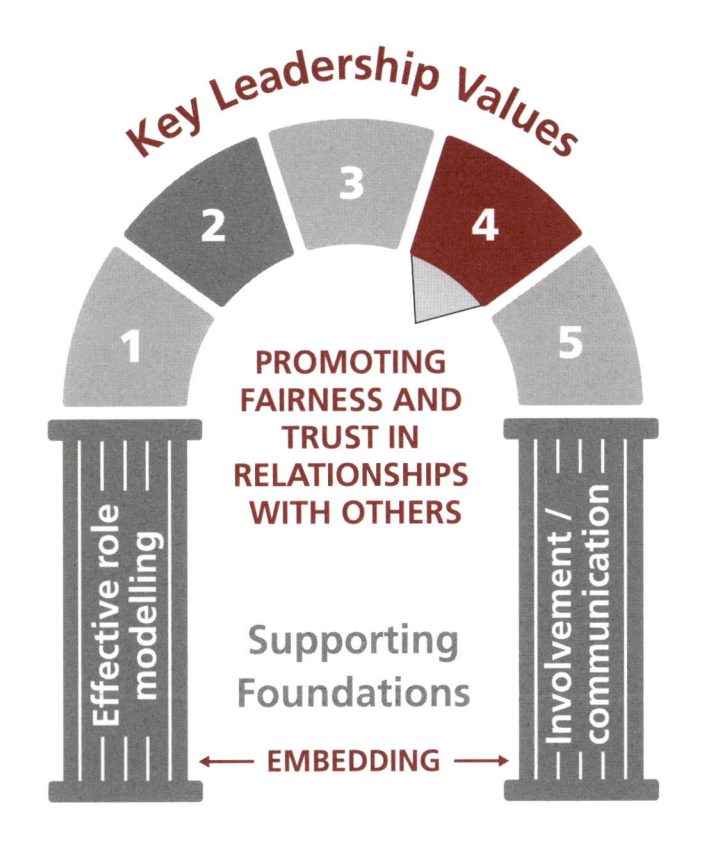

Key Leadership Values

4 — PROMOTING FAIRNESS AND TRUST IN RELATIONSHIPS WITH OTHERS

Supporting Foundations

← EMBEDDING →

Effective role modelling | Involvement / communication

Assessment Activity
- Please refer to the document Unit HSL1, guidance and information for candidates and internal assessors.

You should now complete task L4: Promoting fairness and trust in relationships with others.

Element 1 The foundations of health and safety leadership

1.4 The influence of good health and safety leadership on health and safety culture

Environmental, health and safety (EHS) and management as a conduit for change

The International Organization for Standardization (ISO) have published standards to help organisations manage their environmental aspects (hazards) and health and safety risks. The environmental standard is ISO 14001:2015 *Environmental Management* and the health and safety standard is ISO 45001:2018 *Occupational Health and Safety Management*. By following these standards, organisations are showing their commitment to protect the environment and the health and safety of their workers and others who visit their site. One of the clauses of a management system relates to improvement (including continuous improvement). Organisations should be striving continuously to improve their own performance and, where possible, to influence change and improvement within other organisations.

One way of influencing other organisations, is via the supply chain. Organisations with mature, well managed EHS management systems can insist that these higher standards are cascaded down through their supply chains. As organisations adopt higher EHS standards, they will very often require the same standards from other organisations that they work with. This can go both ways within a supply chain, that is, not purchasing from companies with poor EHS standards, as well as not selling products to them. Very often supply-chain audits are carried out before contracts are signed to ensure that businesses are actually doing what they say they are doing. The possible loss of business is a big incentive for organisations to change their health and safety culture and standards.

Larger organisations will very often publish details about their Corporate Social Responsibility (CSR) activities. These organisations will want to be seen as exemplars within their industry and will go beyond the minimum legal standards. This could, in turn, have an effect on competitors. The competitor will not want to be seen as less ethical than their competition and will, therefore, raise their own standards. This can also influence their own supply chains.

Some organisations, however, may not need the threat of lost business to change; some companies will change for ethical reasons; they want to be seen to be doing the right thing. This is very important to many businesses, especially with the influence of the media, social media in particular. It only takes one story about unethical trading to go viral to ruin the reputation of a business.

Blame culture, no name no blame and just culture

Blame culture

A blame culture can often be the default culture because of the inherent human need to shift blame to others.

An organisation that has a blame culture seeks to find out who is responsible, then to attribute blame. This could result in consequences including disciplinary action, dismissal and retraining. In a blame culture, it would also usually lead to the blame for the incident being placed on an individual or group. This culture, with a focus on blaming people, does not encourage the sharing of information on the actions that led to the errors because people do not want to be blamed. In this culture, people are unlikely to share knowledge or report incidents as they are afraid of recriminations; additionally, there will be a lack of organisational learning.

It is, therefore, unlikely that a blame culture will lead to long-term improvements to health and safety. However, blame culture sometimes seems prolific; this may be because blaming people can provide a simple solution and is less complicated and time consuming than properly investigating the issue and reviewing and changing processes and procedures.

1.4 The influence of good health and safety leadership on health and safety culture

Workers within an organisation can believe that their organisation has a blame culture (default setting) when, in fact, this is not the case. However, just the belief in an organisational blame culture can be enough to stop workers reporting accidents and near misses. It is, therefore, important that health and safety leaders demonstrate a commitment to an alternative culture to change the organisation's default setting of blame culture.

> **Activity**
>
> Why do you think a blame culture may develop within an organisation?
>
> What might some indicators be and how would you address them?

No name, no blame

The opposite of the blame culture is a no name, no blame culture.

This culture is built on the belief that blame is not the issue. If a problem occurs, it is investigated to try and find out why it happened. Workers are encouraged to speak openly about problems and mistakes. The workforce and managers are empowered to be honest and open; the most vital consideration is making things work properly and preventing mistakes from happening again.

However the issue with no name, no blame is that nobody is held accountable for their actions.

Just culture

A just culture takes the no name, no blame idea one stage further to a 'just, no blame culture' or 'just culture'. A just culture is the balancing point between a blame culture and a no name, no blame culture.

Here, blame would only be attributed when somebody has been reckless and taken unnecessary and unacceptable risks. An effective health and safety leader will be required to show support for workers when things genuinely go wrong. This will encourage the reporting of incidents without fear of blame. This, in turn, will mean the organisation will benefit from understanding why accidents or incidents occurred and how to take appropriate action to prevent them happening again.

However, in a just culture leaders must also recognise that they must give recognition for safe behaviour. They must also appreciate that accountability must be delivered in a fair and consistent manner, no matter what level the individual is at within the organisation. Consistent accountability and recognition are two things that many organisations do badly.

Element 1 The foundations of health and safety leadership 39

1.4 The influence of good health and safety leadership on health and safety culture

Three-aspect approach to health and safety culture

The following diagram is from the HSE's *Research Report 367*[26] and shows the three-aspect approach to Health and Safety Culture.

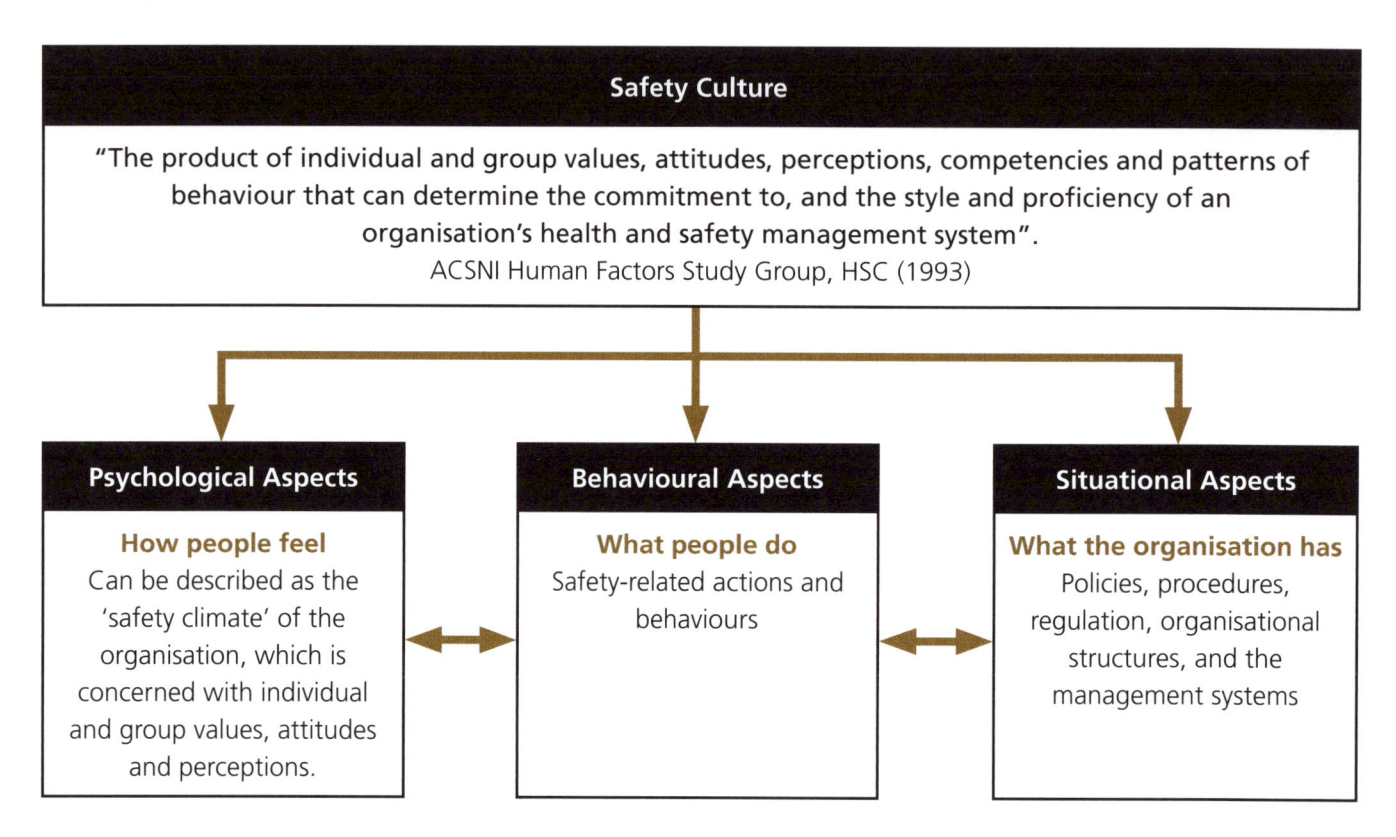

Safety Culture

"The product of individual and group values, attitudes, perceptions, competencies and patterns of behaviour that can determine the commitment to, and the style and proficiency of an organisation's health and safety management system".
ACSNI Human Factors Study Group, HSC (1993)

Psychological Aspects
How people feel
Can be described as the 'safety climate' of the organisation, which is concerned with individual and group values, attitudes and perceptions.

Behavioural Aspects
What people do
Safety-related actions and behaviours

Situational Aspects
What the organisation has
Policies, procedures, regulation, organisational structures, and the management systems

The three-aspect approach to health and safety culture

The diagram is based on the work of Cooper (2000).[27] This illustrates that an organisation's safety culture will depend on three aspects: psychological, behavioural and situational. The connecting arrows show that the three aspects are interrelated and not standalone elements.

Psychological Aspects are about how people feel within the organisation and why. This may have more to do with the workforce's perceptions of how the organisation may ensure a safe and secure environment. It will also include individuals' attitudes about the organisation, each other, the work they do, the environment in which they work, as well as their own beliefs and values built over time. Attitude affects behaviour.

Behavioural Aspects are how people (workforce, management, leaders) act with regard to health and safety within the workplace; they also take account of the impact that people's behaviour has on the overall safety culture of the organisation.

Situational Aspects comprise the organisation's own processes, rules, regulations and procedures as well as how it is set up and what the culture looks like. These aspects determine the messages being communicated to the wider workforce about the organisation's attitude towards health and safety.

Together, these three factors form the basis of the safety culture of an organisation. If individual members or groups within the workforce feel there is a lack of interest in following safety procedures, it could lead to the perception of a poor safety culture; this, in turn, could promote negative behaviour. It is, therefore, vital for an effective health and safety leader to understand the importance of monitoring and maintaining the organisational aspects of the culture. The leader will also need to keep a close eye on workforce attitudes by engaging in a continuous conversation with those involved.

1.4 The influence of good health and safety leadership on health and safety culture

Levels of maturity in health and safety culture

It is important for anyone involved in health and safety leadership to be aware of where their organisation is positioned with regard to a health and safety culture. As with any sort of strategy it is essential to know 'where you are now' so that planning will be effective for improvement and growth. The following diagram sets out the characteristics that can be used to illustrate and explain the different stages of maturity an organisation may travel through on the journey from Ad-Hoc to Excellence.

HSE's levels of cultural maturity

SAFETY/BUSINESS INTEGRATION - WORKER INVOLVEMENT

LEVEL 1 AD HOC
LEVEL 2 MANAGED
LEVEL 3 STANDARDISED
LEVEL 4 PREDICTABLE
LEVEL 5 EXCELLENCE

LEARNING AND ANTICIPATION - LEADERSHIP - SAFETY RESPONSIBILITY

Level 1: Ad Hoc	- Safety is not considered to contribute to business success and is perceived to be a burdensome addition to day-to-day work. - Safety is seen as the responsibility of the health and safety department by everybody in the organisation. - Senior leaders are willing to compromise safety if it appears to be a barrier to productivity and where workers have little interest in safety. - Accidents are seen as unavoidable and if investigations are undertaken the focus is on who is to blame.
Level 2: Managed	- Safety is recognised as contributing to business success but is seen as an additional element of day-to-day working. - Managers recognise that they have a role to play in safety but the health and safety department take the lead. - Individuals recognise that they have responsibility for their own health and safety and are occasionally consulted about health and safety. - Accidents are seen as avoidable and investigations are undertaken but are limited in scope, response or learning. - Leading indicators for safety are not considered.

Element 1 The foundations of health and safety leadership

1.4 The influence of good health and safety leadership on health and safety culture

Level 3: Standardised	- Safety is understood to contribute to business success and it is common practice to consider it in day-to-day work. - Everybody recognises that they contribute to health and safety but the health and safety department is regarded as the lead and workers tend to look out for themselves and their immediate colleagues. - Senior leaders are committed to health and safety but this commitment is not always realised at all levels of the organisation. - Workers are willing to take part in consultation but this tends to be reactive and piecemeal. - There is an acknowledgement that organisational factors can contribute to accidents and there is some learning from investigations. - A limited range of leading indicators are monitored to identify potential safety issues.
Level 4: Predictable	- Safety is routinely considered when making business decisions and in day-to-day working. - Everybody recognises that they are responsible not only for their own safety but also the safety of their colleagues throughout the organisation. - There is a clear commitment to health and safety by leaders at every level. - Engagement with workers is proactive and meaningful. - Lessons are learnt from both accident and near miss investigations. - There are a good range of leading indicators that are regularly monitored to identify health and safety issues.
Level 5: Excellent	- Safety is considered to be critical to business success and is a valued, integral element of day-to-day work activities. - Everybody within the organisation recognises that they are responsible for safety. - There is routine, visible senior leadership and strong partnership working with all levels within the organisation. - Accident and near miss investigations consider the full range of root causes and the emphasis is learning from accidents and near misses. - The organisation actively exploits a wide variety of information to anticipate potential safety issues.

It should be noted, however, that there could be several different levels of cultural maturity across an organisation. For example, an organisation that has multiple sites or operates in different locations around the world.

Activity

From the characteristics we have just looked at, where do think your own organisation sits within these levels?

What sort of evidence would you present to support your argument?

Where, realistically, does your organisation aspire to be. What is your end point and why? Is compliance (mid-point) enough? Will reaching the top of the scale involve disproportionate investment?

1.4 The influence of good health and safety leadership on health and safety culture

Leading and lagging indicators of health and safety culture

> **Key Terms**
>
> **Indicators (leading and lagging)**
> Measurements taken to assess safety performance and determine what needs to be done to improve the safety culture of an organisation.
>
> **Leading indicators**
> Leading indicators are proactive; they aim to prevent adverse events before they happen.
>
> **Lagging indicators**
> Lagging indicators are reactive; they aim to measure the effectiveness of safety management systems after events have occurred.

It is important for a health and safety leader to understand that there are a range of indicators that can be used to help them manage health and safety. Leading indicators can help to improve safety culture by preventing unwanted events from happening. Having a good range of leading indicators this will show workers and other relevant stakeholders that the organisation is taking health and safety seriously. Lagging indicators are useful in identifying trends but, as they measure past events (normally failures), they are not particularly useful when trying to prevent unwanted events from happening.

Leading indicators that can be used include:
- results of safety campaigns;
- safety training for the workforce;
- number of toolbox talks;
- risk assessments;
- outcomes from health and safety committees/worker involvement in health and safety;
- lessons learnt databases;
- job safety analyses;
- leadership site visits/number of walkabouts;
- review of procedures (completed on time and have involved relevant workers);
- suggestion programmes and evidence of changes made;
- training opportunities, assessments and feedback;
- number of production down-times (if related to safety);
- safety audits; and
- planned maintenance programmes for the workplace and work equipment/machinery.

Note: it is important to bear in mind not just the quantity of these metrics, but also the quality. For example, how competence is assessed following training.

Lagging indicators that can be used include:
- number of fatalities;
- number of days lost through ill-health or workplace injuries;
- number and types of ill-health and injuries;
- how often accidents may happen;
- number of prosecutions or civil actions;
- number of enforcement notices issued by regulators; and
- number of reportable incidents.

Element 1 The foundations of health and safety leadership

1.4 The influence of good health and safety leadership on health and safety culture

Measuring the right things

For any leader to make progress, it is important to understand where an organisation sits on the scale of health and safety cultural maturity. To enable this, the organisation must measure and monitor procedures and processes. However, to have any real value, these must be the right procedures and processes. It is surprising, but all too common for organisations to measure things which are easy to find or show things in a favourable light. This is not at all useful for the leader who really wants to create an effective health and safety culture.

A well-used model of measuring risk is James Reason's Swiss Cheese Failure Model. Reason compares defensive layers to several layers of Swiss cheese. Each layer of cheese represents a defence against a hazard being realised and each defence has holes in it.

The model recognises that things do go wrong from time to time and that something breaking through one of the holes is not too much of a problem. Minor failures can be dealt with as part of normal routine in which errors are fixed and things move on.

However, things tend to go wrong when holes align through all of the layers. This indicates that all risk control systems have failed and there will be a major failure or the hazard is realised.

The following diagram gives an example of how the holes easily align due to the limited amount of checking which will cause failure.

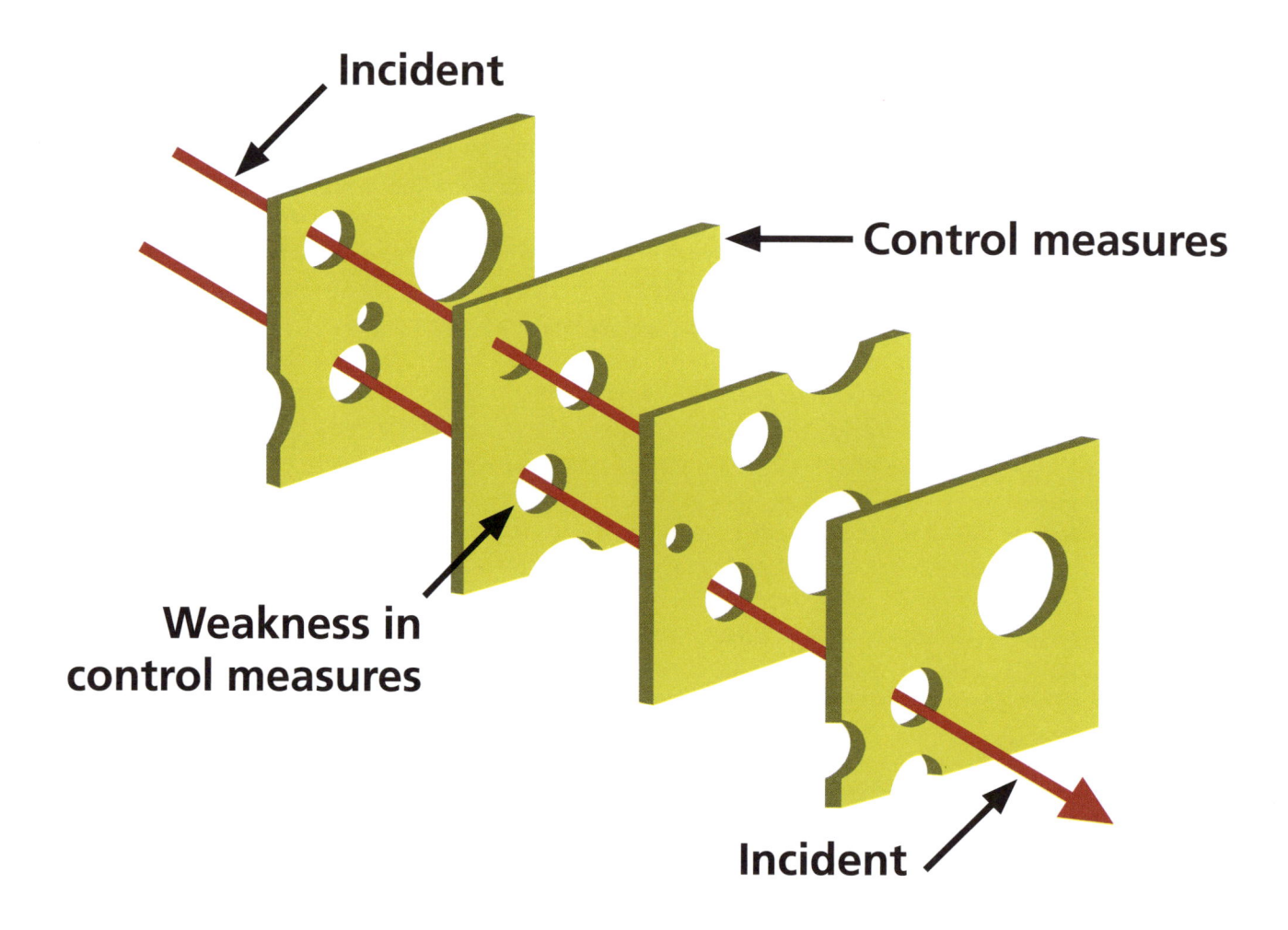

Putting in place extra checks will make it more difficult for the hole to align. This reduces the chances of the hazard being realised.

Element 1 The foundations of health and safety leadership

1.4 The influence of good health and safety leadership on health and safety culture

The Swiss Cheese Model has been used extensively in risk management, health care, aviation, and engineering. It is very useful as a method for explaining the concept of cumulative effects.

Models like this are essential tools in the toolkit of the effective health and safety leader. As is learning from other organisational practices and benchmarking against any available best practice. The industry or sector from which this best practice comes is often irrelevant; with slight modification, a process, procedure or working practice can be rewritten to fit the needs of an organisation.

High Reliability Organisations (HROs)

Key Term

High Reliability Organisation (HRO)
This is an organisation that "is able to manage and sustain, almost error-free, performance despite operating in hazardous conditions where the consequences of errors could be catastrophic".[28]

HROs were first developed in the following sectors:
- power grid dispatching centres;
- air traffic control systems;
- nuclear power plants;
- wildfire fire-fighting;
- aircraft operations; and
- accident investigation teams.

HROs have capacity to maintain or regain a stable state which is something all organisations should aspire to. The following diagram, from HSE's *Research Report 899*, illustrates the characteristics generally found in HROs.[29]

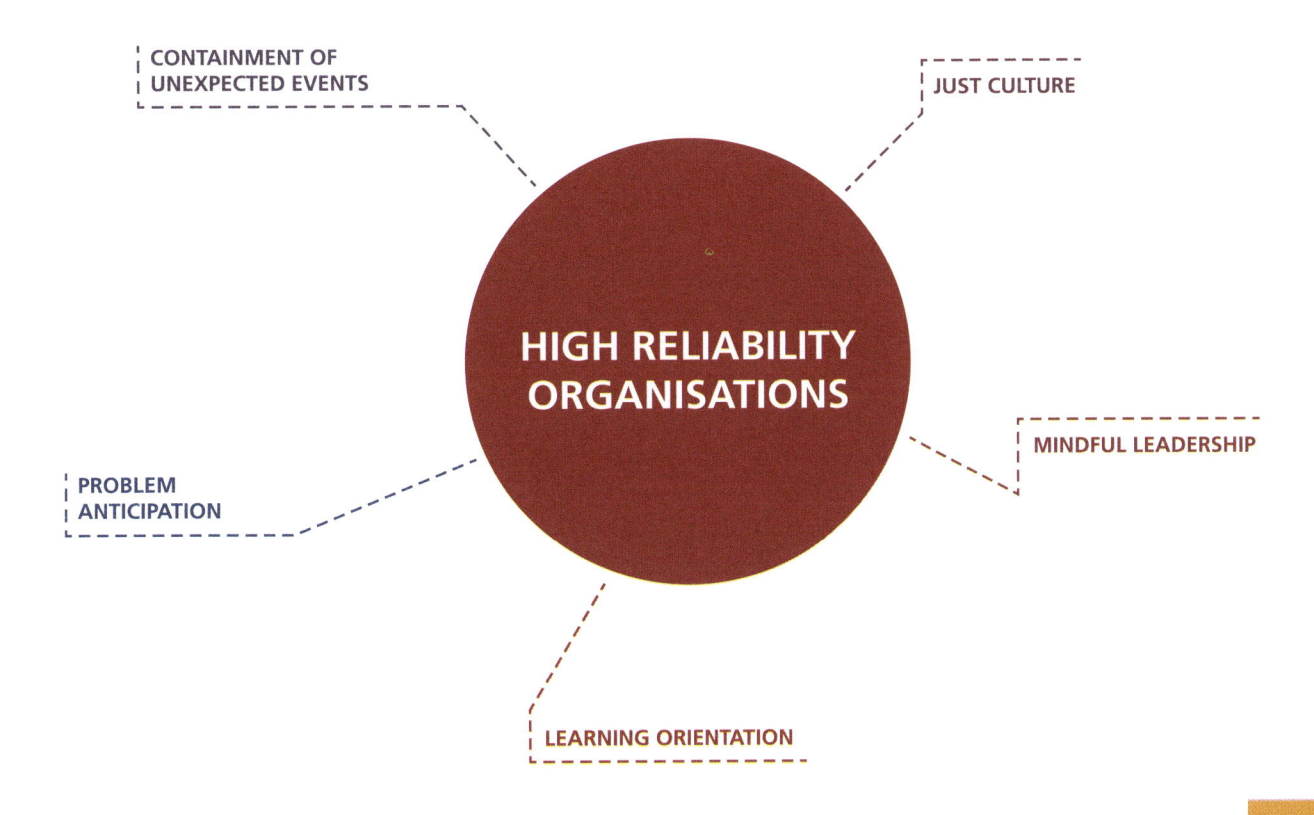

HIGH RELIABILITY ORGANISATIONS
- CONTAINMENT OF UNEXPECTED EVENTS
- JUST CULTURE
- MINDFUL LEADERSHIP
- LEARNING ORIENTATION
- PROBLEM ANTICIPATION

Element 1 The foundations of health and safety leadership

1.4 The influence of good health and safety leadership on health and safety culture

Containment of unexpected (emergency) events

HROs will generally:
- invest heavily in technical expertise within their organisation. They will value those with such expertise and listen to and follow their advice/instructions;
- allow their experts to make important safety-related decisions in emergencies; during normal operating conditions there is a clear hierarchical structure and an understanding of who does what;
- invest in training for workers at all levels to ensure that the organisation has the right levels of competence throughout the organisation; and
- will ensure that they have back-up/redundancy systems in place and have emergency plans in place for unexpected events.

Just culture

A just culture is one where accountability and recognition are dealt with consistently throughout the organisation.

Mindful leadership

HROs generally will have leaders who are:
- visible and actively engage with the workforce;
- prepared to receive bottom-up communications which have bad news; they will ensure that there are communication channels available to allow this type of reporting;
- able to ensure that sufficient resources are made available for all operations; whether this be people or equipment;
- able to balance profits with safety for example, by encouraging all staff to follow procedures and not cut corners; and
- involved in proactively commissioning audits to identify problems in systems; this can be in response to incidents that occur in other similar industries.

Learning orientation

HROs will general have systems in place which allow for:
- continuous technical training;
- systematic analysis of incidents to identify their root causes and accident types or trends within the organisation;
- open communication of accident investigation outcomes; and
- updating procedures in line with the organisational knowledge base.

Problem anticipation

HROs will also generally be able to anticipate potential failures by:
- engaging with front line workers in order to obtain an overview of operations (sensitivity to operations);
- attentiveness to minor, or what may appear as trivial, signals that may indicate potential problem areas within the organisation and using incidents and near misses as indicators of a system's health (preoccupation with failure); and
- carrying out systematic collection and analysis of all warning signals, no matter how trivial they may appear to be, and avoiding making assumptions regarding the nature of failures. Explanations regarding the causes of incidents tend to be systemic rather than focusing on individuals.

1.4 The influence of good health and safety leadership on health and safety culture

Over-reliance on technology

One of the limitations of HROs is their over-reliance on technology to avoid unexpected events. Generally these organisations tend to maintain back-up and redundancy systems to help manage these events.

The problem with this is that there is the possibility of relying too much on technology and ignoring the human inputs.

> **Activity**
>
> Do you recognise any of the HRO characteristics within your own organisation?
>
> How close do you feel you are to this type of health and safety culture?
>
> How would you, as a leader, influence the organisation to move in this direction?

Element 1.4 References

23 Confederation of British Industry, *Developing a Safety Culture: Business for Safety* (Confederation of British Industry, 1990)
24 Handy, C, *Understanding Organisations* (4th edition, 1993, Penguin)
25 HSE, *HSE Human Factors Briefing Note No 7. Safety Culture* (2003). See also HSE, 'Human factors in risk assessment' (www.hse.gov.uk/humanfactors/resources/risk-assessment.htm)
26 HSE, *A review of the safety culture and safety climate literature for the development of the safety culture inspection toolkit* (RR367, 2005) (www.hse.gov.uk/Research/rrhtm/rr367.htm)
27 Cooper, MD, 'Towards a Model of Safety Culture' (2000) *Safety Science* 36(2), 111-136
28 HSE, *High reliability organisations – A review of the literature* (RR899, 2011) (https://www.hse.gov.uk/Research/rrhtm/rr899.htm)
29 RR899, pages 17-18. See note 28.

Element 1 The foundations of health and safety leadership

Element 2
Human failure and decision making

This Element will explore how human failure can impact the health and safety culture of an organisation. It will then look how we all use mental short cuts, biases, habits and beliefs to influence our decision making process.

Learning outcomes
- Recognise the impact of human failure on health and safety culture.
- Understand the impact of mental short cuts, biases, habits and beliefs on the decision making process.

2.1 The influence of human failure on health and safety culture

This section examines the relationship between human failure and decision making and how human failure impacts safety culture.

> **Key Terms**
>
> **Error:** an unintentional action or decision.
>
> **Violation:** an intentional failure or deliberately doing the wrong thing.

The following illustration is from the HSE's publication *Reducing error and influencing behaviour* (HSG48)[30] and illustrates the various types of human failures.

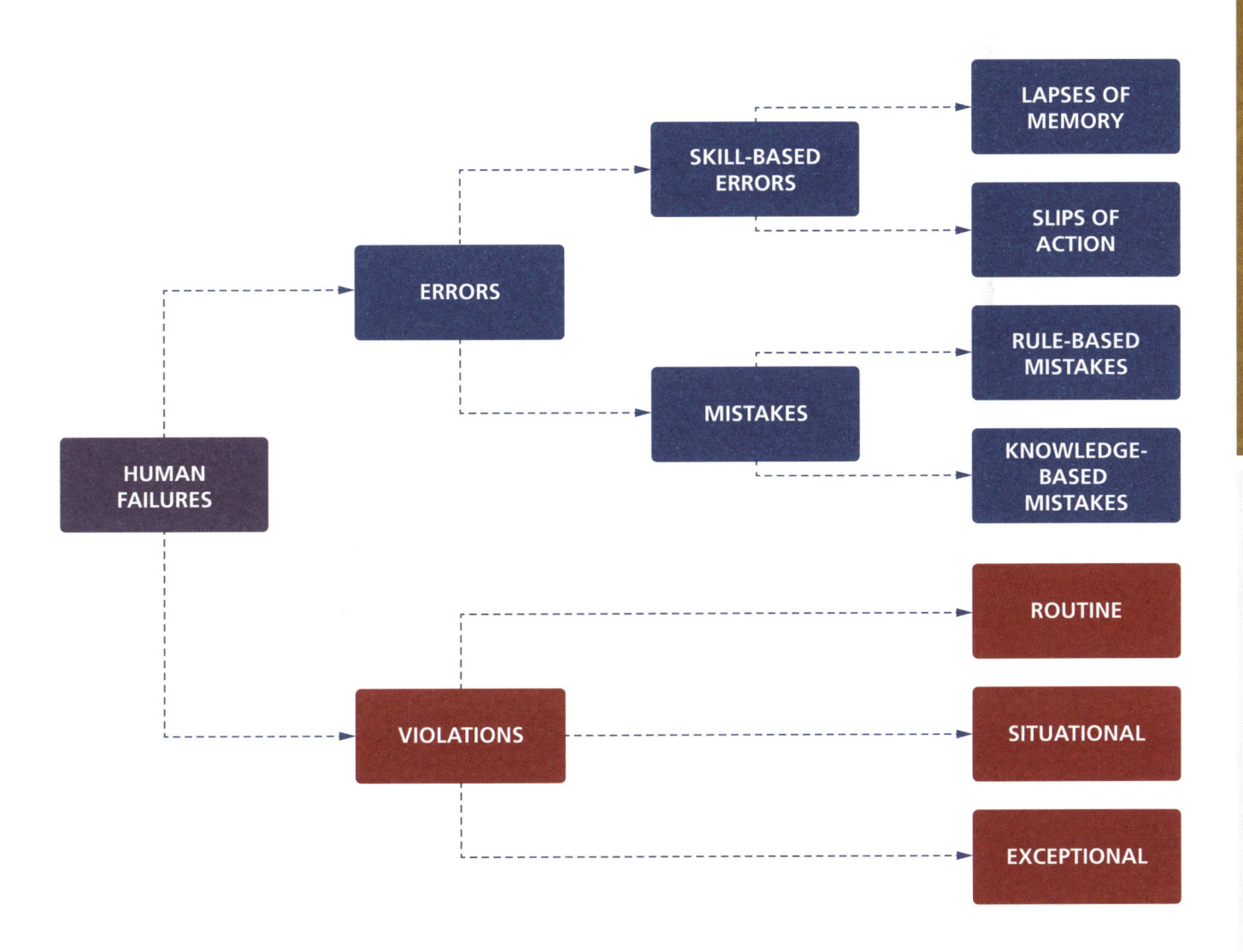

```
HUMAN FAILURES
├── ERRORS
│   ├── SKILL-BASED ERRORS
│   │   ├── LAPSES OF MEMORY
│   │   └── SLIPS OF ACTION
│   └── MISTAKES
│       ├── RULE-BASED MISTAKES
│       └── KNOWLEDGE-BASED MISTAKES
└── VIOLATIONS
    ├── ROUTINE
    ├── SITUATIONAL
    └── EXCEPTIONAL
```

Source HSE, *Reducing error and influencing behaviour* (HSG48)

Element 2 Human failure and decision making

2.1 The influence of human failure on health and safety culture

Errors

Skill-based errors

Errors usually occur in very familiar tasks the person does not need to pay much attention to what they are doing. One of the most common examples of a skill-based activity is driving a car. If a slip of action or a lapse of memory occurs, even momentarily, the consequences can be devastating. Slips and lapses can happen even to well-trained, experienced individuals.

Skill-based errors are broken down into 'slips of action' and 'lapses of memory'.

Slips of action

These are unintentional failures to carry out all actions of a task, in other words, unplanned actions. Some examples of slips of action are:
- performing an action either too soon or too late in a procedure;
- forgetting to perform one or more steps in a procedure;
- carrying out the action in the wrong direction, for example, turning a piece of machinery to a higher speed instead of turning it off;
- performing the right activity but on the wrong piece of equipment, for example, flicking a switch on or off but not on the right piece of equipment; or
- carrying out the wrong check but on the right piece of equipment, for example, checking a dial but recording the
wrong information.

Lapses of memory

Again these are unintentional. A lapse means that a person forgets to carry out an action in activity, or forgets where they are in a particular process, meaning they could do the right thing but at the wrong time.

Lapses of memory can be minimised by:
- minimising distractions, for example, by trying to stop workers from talking to each other while performing an activity;
- providing reminders of the steps involved in the activity, for example, providing a checklist that can be completed after each stage of an activity;
- designing tasks in a better way, for example, so that the operator does not need to leave their workstation half-way through an activity.

2.1 The influence of human failure on health and safety culture

Mistakes

Mistakes happen when a person thinks that they are doing the right thing but it is actually wrong. Mistakes are a more complex form of errors.

Mistakes are broken down between 'rule-based mistakes' and 'knowledge-based mistakes'.

Rule-based mistakes

A person's behaviour is based on remembered rules or familiar procedures. There is a strong inclination to use the familiar even when this may be wrong or not the most efficient way of dealing with the situation. An example of a rule-based mistake is where a tanker driver ignores high level alarms when filling a tank because he knows how long the tank takes to fill as he has done it many times before. However, on this occasion he does not realise that the diameter of the pipe entering the tank has been enlarged meaning that the tank fills more quickly.

Knowledge-based mistakes

These mistakes tend to happen in unfamiliar situations and there are no 'tried and tested' rules to rely upon. A person tends to rely on knowledge of similar situations to try to reach a solution for the unfamiliar situation. An incorrect decision is made because an individual doesn't have all the information they need, or they simplify the available information and make a decision too quickly without considering all the options. Sometimes these errors occur at work because an individual has to deal with a situation that is beyond their level of understanding.

An example of a knowledge-based mistake is when someone tries to amend a process by relying on information they did not know was out of date or from their past experiences.

Violations

Reducing error and influencing behaviour (HSG48) refers to violations as deliberate deviations from rules, procedures and instructions designed for safety and efficiency of a system.

Slips and lapses may be attributed to human error or simply because humans make mistakes. Violations are a deliberate attempt to avoid or ignore procedures and instructions that have been designed to safeguard people and property. The reasons for violation of these processes and deliberately doing the wrong thing may be many and varied. They will however, according to the HSE, fall into one of three categories; routine, situational or exceptional. The HSE report *Improving compliance with safety procedures, reducing industrial violations*[31] outlines practical strategies for reducing the potential for violations, as well as highlighting some significant and devastating examples of violations and their consequences.

Routine violations

A routine violation is where breaking the rules has become a normal way of working within the workforce. This can be for different reasons, including:
- cutting corners to save time;
- the workforce may feel that the rules are too prescriptive;
- if carried on for a long period of time, the workforce may feel that the rules no longer apply;
- if leadership does not enforce rules, the workforce will continue to ignore them; and
- new workers entering the workforce may not realise that routine violations are the norm; they may not realise that rules apply.

Element 2 Human failure and decision making

2.1 The influence of human failure on health and safety culture

Situational violations

A situational violation occurs in response to situational factors, including excessive time pressure, workplace design, extreme weather conditions and inadequate or inappropriate equipment. When confronted with an unexpected or inappropriate situation, workers may believe that the normal rule is no longer safe, or that it will not achieve the desired outcome, and so they decide to violate that rule. Situational violations generally occur as a one-off, unless the situation triggering the violation is not corrected, in which case the violation may become routine over time.

Exceptional violations

An exceptional violation is a fairly rare occurrence and happens in abnormal and emergency situations. This type of violation transpires when something is going wrong and workers believe that they must break the rules even though they know this is a risk. Workers choose to violate the rule believing that they will achieve the desired outcome.

Impact on safety culture

The safety culture of an organisation influences how individuals handle new events and decisions. If the wrong messages about health and safety are received, this may be seen to encourage rule breaking. Similarly, the type and frequency of human failures seen in an organisation, together with management's response to those human failures, may impact the safety culture of that organisation. Repeated routine violations, for example, may be seen as the norm which could have a negative impact on safety culture.

2.1 The influence of human failure on health and safety culture

Providing support and recognition (health and safety leadership value 3)

People recognise when they are valued, and they should be provided with a good idea of their value to the organisation. If workers are treated as valued team members, not as numbers, they will respond positively. When a leader shows genuine interest in supporting others in the workforce there is a positive effect.

Training and development are key to ensuring workers have the right tools to carry out the job. Identifying what is needed requires a continuing dialogue. Leaders should ensure that advice and guidance concerning best practice in health and safety is readily available to anyone who looks for it. Importantly, this advice and guidance should be assessed to ensure that it is accessible and the language is appropriate for the audience. There is little point in distributing training or guidance if the users have difficulty finding or using it. This can also cause frustration and may lead to disengagement and scepticism.

Recognising and rewarding the efforts made by individuals and teams will help motivate and encourage workers to continue their efforts. Rewards do not have to be financial; they can come in many different forms.

People are motivated by many different things. These can include being singled out for recognition and providing extra training opportunities for those who aspire to improve their skills. Providing extra training can also be very effective in building strong long tem relationships. Providing resources fairly and without prejudice where and when they are needed shows that support is real and ongoing. This builds trust.

Assessment Activity
- Please refer to the document Unit HSL1, guidance and information for candidates and internal assessors.

You should now complete task L3: Providing support and recognition.

Element 2.1 References

30   HSE, *Reducing error and influencing behaviour* (HSG48, 2nd edition, 1999) (https://www.hse.gov.uk/pubns/books/hsg48.htm)
31   HSE Human Factors in Reliability Group, *Improving compliance with safety procedures. Reducing industrial violations* (1995, HSE Books) (www.hse.gov.uk/humanfactors/assets/docs/improvecompliance.pdf)

2.2 Decision-making processes, mental shortcuts, biases and habits

The differences between 'Automatic' and 'Reflective' decision making

Some decision-making processes will have an influence on behaviour. Understanding different types of decision-making in more detail and the differences between Automatic and Reflective decision-making can begin by looking at how humans make decisions. While there are many and varied influences and reasons behind why someone chooses to do what they do, people generally approach different types of decision in different ways.

For example, decisions about which area to live in, which property to live in, or which career to pursue are, for most people, important decisions, requiring care and attention and unhurried, considered responses. They may have a gut feeling about which to choose but are still likely to carry out research, review options, and seek the opinions of others to arrive at a rational decision based on investigation and a period of reflection. This is a 'reflective' decision-making process.

People are unlikely to adopt a similar approach to buying a chocolate bar, choosing which TV programme to watch or which seat to occupy on a train. While carrying out detailed research and building a list of pros and cons about houses and careers is a sensible and appropriate response, doing the same for confectionary, a train ride or other seemingly unimportant but necessary daily decisions would waste huge amounts of time and achieve very little. This is why people often rely to a considerable extent on 'automatic' decision-making for these types of tasks.

This method enables quick decisions to be made, based on limited information but supplemented by a variety of 'shortcuts' that are developed during a person's life. Marketers know this very well and use it with extraordinary effectiveness to ensure people choose their particular product or service over their competitors without ever really knowing why.

These two methods of making decisions are present in everyone. In his book *Thinking Fast and Slow*,[32] psychologist Daniel Kahneman describes these two systems in the mind as System 1 and System 2:
- System 1 is the automatic, rapid almost unconscious way of thinking which is efficient, requires little or no effort or attention, but is prone to slip-ups and mistakes.
- System 2 is reflective and requires effort and attention, it is a slower, more controlled way of thinking. It also has the capability to moderate the instincts present in System 1.

The human brain can make split-second decisions without any effort on our part at all and we have no voluntary control over this aspect of our brain. This system is helpful in certain situations, but it can also cause trouble because it can be emotional. System 1 is biased to believe and confirm. It also focuses on existing evidence and ignores any lack of evidence. Kahneman calls this "what you see is all there is".

The following diagram shows Kahneman's framework for fast and slow thinking.

2.2 Decision-making processes, mental shortcuts, biases and habits

	SYSTEM 1	SYSTEM 2
CHARACTERISTICS	Fast Effortless Unconscious Triggers emotions Associative Looks for patterns Looks for causation Creates stories to explain events	Slow Effortful Conscious Logical Deliberative Can handle abstract concepts
ADVANTAGES	Speed of response in a crisis Easy completion of routine or repetitive tasks Creativity through associations, so is good for expansive thinking	Allows reflection and consideration of the "bigger picture", options, pros and cons, consequences Can handle logic, maths, statistics Good for reductive thinking
DISADVANTAGES	Jumps to conclusions Unhelpful emotional responses Can make errors that are not detected and corrected, such as wrong assumptions, poor judgements, false casual links	Slow, so requires time Requires effort and energy, which can lead to decision fatigue

The human attention span has a limited capacity and the way in which we act reflects this.

When instructed to focus on one task, people can become 'blind' to other activities going on in the same area. Christopher Chabris and Daniel Simons illustrate this in the book *The Invisible Gorilla*.[33] They produced a short video of two teams playing basketball; one team wearing white tops and the other wearing black tops. The instruction at the beginning of the video is to count the number of passes made by the team wearing the white tops (so totally discounting the team in black). At the end of the video people are asked how many passes they counted. They are then asked, 'did you see the gorilla?'. Over half of the people who watch the video for the first time miss the gorilla, although it stays in the video for roughly 10 seconds.

System 1 thinking is used to carry out this observational task; to be able to see the gorilla, people need to be able to notice unexpected stimuli in a situation. When a person cannot concentrate on all stimuli, they can become temporarily blind to stimuli that are additional to the observational task and ignore them.

Element 2 Human failure and decision making

2.2 Decision-making processes, mental shortcuts, biases and habits

Reliable mental shortcuts

Some estimates suggest that a person can make up to 32,000 decisions in a single day, many of which they do not even know they are making. It is no surprise that humans have developed a range of mental shortcuts to help (or hinder) the process of making decisions.

> **Key Term**
>
> **Heuristics:** Simple rules or mental shortcuts that people use to form judgments and make decisions.

Where organisational risks are well known, heuristics can be effective when estimating the level of risk. The draw back to this is that we can sometimes place too much belief in judgements reached by using heuristics.

Understanding how heuristics affect decisions is critical in developing learning and response in the assessment and management of risk and safety.

Here are some of the more common mental shortcuts used in decision making:
- Anchoring
- Availability
- Representativeness
- Media influences

Anchoring

The anchoring shortcut (or rule of thumb) describes the common human tendency to rely too heavily on the first piece of information offered (the anchor) when making decisions. Once the anchor is set, decisions are then made by adjusting around the initial anchor, regardless of the legitimacy of the actual anchor.

For example, when making a decision about whether to purchase an item, a person needs an initial figure which provides a focal point such as a recommended retail price. If this is given as £600 but the person thinks that this is too high, they are far more likely to buy the item at £150, even if it was never worth £600 and its real value was £100. The person has anchored to a number and this has become part of their decision-making process.

In a health and safety scenario anchoring may act as a reference point that connects a historical event with the present and uses past experiences to influence decisions. For example, if a supervisor was involved in a serious forklift truck incident at some stage in the past, further discussion of this topic may trigger an anchoring response based on this past experience. This may result in either a raised level of awareness and knowledge, or conversely, perhaps a degree of over-sensitivity and a reluctance to engage.

The effective health and safety leader will need to understand the situation and consider whether a more sensitive line of enquiry would be better rather than using direct questioning techniques.

2.2 Decision-making processes, mental shortcuts, biases and habits

Availability

Recalling some things more easily than other things is normal for most people. Availability shortcuts assist in estimating how likely something is to happen based on information a person can. People make judgements about the likelihood of an event or situation based on how easily an example comes to mind.

The availability heuristic can be helpful when someone is faced with a choice but they do not have the time to investigate before making their final choice. The availability heuristic will help that person reach an immediate decision. However, this could result in an incorrect decision.

For example, if a person hears a traffic update at the same time each day featuring a particular road, they may conclude that this is a particularly troubling route. But, this may not be true; it is just the road that they know about. Similarly, in the workplace, if someone is exposed to several incidents that took place in a specific area, they may decide that this is a particularly difficult area and requires special consideration. This may be right, but it may also be because they recall more about this area, maybe because they has a friend who works there.

The effective leader will understand that these factors require consideration in order to make rational decisions.

Representativeness

This is a type of mental shortcut that enables quick decision making based on previous experience when faced with uncertainty. The representativeness heuristic is simply described as assessing the similarity of objects or events and organising them based around the category prototype so that like goes with like and causes and effects should resemble each other.

The representativeness heuristic is used to judge the probability that object or event A belongs to class B, simply by looking at the degree to which A resembles B.

When people rely on representativeness to make judgements, they are likely to judge wrongly because the fact that something closely resembles does not actually make it more likely.

This mental shortcut may provide quick decisions but will these be the right decisions? The decisions may be based on fallacies, by relying on invalid reasoning. For example, a person who is facing a challenge may rely on an example from the past without really understanding that the circumstances are different and that more information is needed for the comparison to be useful.

A good example of representativeness is called the 'gambler's fallacy' which is where someone believes that runs of good and bad luck occur. So, if a coin toss turns up heads multiple times in a row, people think that tails is a more likely occurrence in the next toss. They think that this will even things out, even though the opposite may happen and that heads will continue the winning streak. This view is held despite the fact that each toss of the coin is a totally independent event and not connected to the one before or after it.

2.2 Decision-making processes, mental shortcuts, biases and habits

Media influence

It is all too easy for decisions to be irrationally based on media influence. Alongside the conventional broadcast and print media, social media has now taken its place with exceptional powers of persuasion and the much discussed 'fake news' elements. If someone sees one of their favourite celebrities endorses Brand A, they may immediately believe the brand to be good; it is easy to forget that the celebrity in question has been paid a lot of money to advertise Brand A! Likewise, if people hear a lot of bad news stories about a particular country, they are more likely not to book a holiday to that destination.

The effective leader needs to be able to take a balanced and reflective position despite media influence and develop the skills and tools to be able to help others in the workforce to better analyse and understand how misleading the media might be. Developing critical thinking and helping to develop this in others will prove a useful technique in health and safety leadership.

Activity
When making decisions, consider the way in which the representative heuristic might play a role in your thinking. What might the dangers be in health and safety leadership of too much reliance on this method?

Individual risk perception

Perception is about the way that people interpret the world around them. What one person sees, or perceives, as being relatively safe, another might think is quite dangerous – take parachuting or bungee jumping as examples. A person's individual risk perception is underpinned by their beliefs, attitudes, judgements and feelings, as well as social or cultural dispositions. Individual perception may also be subconscious.

Individual perception can be based on sensory information, from the five senses (hearing, taste, touch, smell and sight), or can be based on experience or familiarity with a situation. Perception is a complex issue, which needs to be considered when deciding how to reduce accidents at work.

For example, if a worker has a form of natural sensory impairment, they will not receive all the information they need to interpret their surroundings and will thus be more likely to have a distorted perception. An example would be a worker with colour vision impairment who may not be able to tell the difference between one coloured cable and another. Perceptions can also be affected by alcohol and drugs (including prescription medication) which may affect the brain's ability to process information.

A person's ability to perceive a hazard, and the risks associated with it can be lowered by tiredness, having to concentrate intensely or by carrying out dull, tedious or repetitive work. Hazards can be hidden, such as radiation, or masked by environmental issues, such as background noise or poor lighting.

Inadequate training will affect individual perception because the worker will have gaps in their knowledge or will assume something is not hazardous when it is. For example, due to insufficient training, they may perceive that the substance they are working with is harmless when it is, in fact, a skin irritant.

2.2 Decision-making processes, mental shortcuts, biases and habits

Experiences also shape individual perceptions. For example, someone may say, 'this is safe – we have done it this way hundreds of times and never had an accident'. Their perception is that the job is safe, but it may not be; they might have been lucky. This is evidenced by comments during accident investigations along the lines of 'I don't know how that happened – we have always done it that way'.

Another issue is how the brain tries to make sense of unfamiliar patterns. If a person sees something that they have not seen before, they may find themselves saying that it looks like X. That is a verbal expression of what their brain is trying to do – match the abstract or unfamiliar with something that is familiar. For example, consider the following pictures. At first sight, what do you see? Different people will see different things in these images. It is all a matter of perception and how your brain makes sense of these unfamiliar drawings.

| Young lady or old lady? | Faces or vase? | Jazz player or face? |

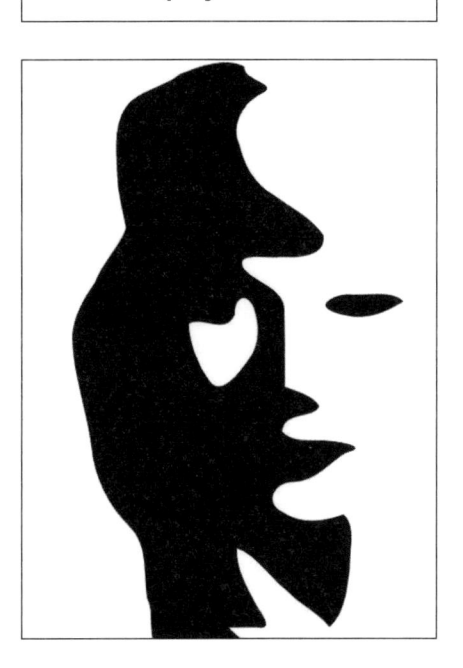

From a health and safety perspective, a person's subconscious perception process can mean that they may not see hazards in the workplace. In fact, people often only see what they expect to see. For example a delivery driver repeatedly driving the same route will get used to the features on the route, such as junctions and traffic lights, so that they no longer perceive them. If there is a change, such as a different speed limit, they may not even notice this.

In summary, different workers may perceive the same thing in entirely different ways. The employer will need to try to ensure that all workers perceive things accurately and that they do not mistake a serious risk for something that is trivial. Increasing the visibility of hazard markings, providing information and training to improve hazard recognition and providing experiences through drills and scenarios can all reinforce the desired perception process and action.

Element 2 Human failure and decision making

2.2 Decision-making processes, mental shortcuts, biases and habits

Common biases and how they affect decision making

Common biases include:
- Halo
- Confirmation
- Self-serving
- Hindsight

> **Key Term**
>
> **Bias**
>
> A strong feeling in favour of or against one group of people, or one side in an argument, often not based on fair judgment.[34]

We all may think we can make fair and accurate judgements or evaluations about situations and events, but can we? The way someone perceives an event or situation is quite often affected by their own biases. Biases can cause someone to feel or show an inclination or prejudice for or against someone or something.

People need to make rapid decisions all the time and to do this, they develop, sometimes unconsciously, methods to quickly arrive at a judgement.

2.2 Decision-making processes, mental shortcuts, biases and habits

Halo effect

The halo effect is a significant bias that, put simply, means that if we find one or two attributes of a person good or attractive, we tend to assume that everything else about them is the same. This can, despite evidence to the contrary, lead to people rating individuals that they perceive as attractive more favourably than those whom they perceive as less attractive. This has been referred to as the 'what is beautiful is good' principle. One good example of the halo effect in action is the way in which people may view celebrities. If a person perceives a celebrity as attractive, successful and likable, they are also more likely to view them as intelligent, kind and funny and this is why marketers use the celebrity halo effect to promote products and services. When a celebrity is used to endorse a product, a person's positive feelings towards the celebrity can positively influence their perception of the product itself.

The halo effect can show up in many areas of life, such as when making a decision about which political candidate to vote for in the next election. A person's overall impression of a particular candidate may influence how they evaluate other characteristics such as how they respond in a crisis.

A workplace example is as follows:
> In the work setting, the halo effect is most likely to show up in a supervisor's appraisal of a subordinate's job performance. In fact, the halo effect is probably the most common bias in performance appraisal. Think about what happens when a supervisor evaluates the performance of a subordinate. The supervisor may give prominence to a single characteristic of the employee, such as enthusiasm, and allow the entire evaluation to be coloured by how he or she judges the employee on that one characteristic. Even though the employee may lack the requisite knowledge or ability to perform the job successfully, if the employee's work shows enthusiasm, the supervisor may very well give him or her a higher performance rating than is justified by knowledge or ability. [35]

It is not enough as a leader to say 'I may not know everything so tell me'; leaders need to take an active role in seeking information.

Activity
Just because we might be aware of the halo effect does not mean we are immune to it. Like heuristics, bias comes from somewhere other than our rational ability to reason. So how can the effective health and safety leader ensure that decisions are made based on more than a feeling?

Confirmation bias

Confirmation bias occurs when decision makers seek out evidence that confirms their previously-held beliefs, while discounting or diminishing the impact of evidence in support of differing conclusions.

Everyone likes to think they are right. How can the effective health and safety leader ensure that decisions are made based on more than 'I told you so'?

Being right is a good sensation. Being wrong however is a totally different feeling and being proven wrong is even worse. This in some way explains why confirmation bias is common. Confirmation bias is based on our desires and pre-existing beliefs; if a person wants something to be true they can eventually end up believing this. However, the issue is that the person stops gathering evidence about the situation, especially when the evidence that has been collected, even if it is very limited, confirms their belief.

Some people may have a selective memory. This is when they choose only to remember, or at least say they remember, only the things they wish to, or that line up with their own beliefs or position.

Most people will, at times, be guilty of confirmation bias and may be unaware of this. Sometimes, we all believe what we want to believe and will interpret uncertain evidence to support our position.

Element 2 Human failure and decision making

2.2 Decision-making processes, mental shortcuts, biases and habits

Good leaders should be aware of confirmation bias. This includes keeping an balanced perspective and being open-minded when undertaking investigations, looking at charts and data and in all conversations. Authentic leaders can recognise confirmation bias.

Self-serving bias

> Well I did work extremely hard and have a natural talent for the subject so I guess that is why I passed so well.

> I don't know how they expected us to pass, the trainer didn't cover half the topics, also because the blinds were not very good, the sun was in my eyes for half of the exam, so I am not surprised I didn't pass.

Consider these two statements. What do you notice about them?

In the first one, which is about success, it is clear that the person attributes their success to personal effort and talent. In the second, which is about failure, it is clearly nothing to do with the individual and the fault lies with other people or conditions. In short this is what self-serving bias is about.

We all need defence mechanisms that will spring into action when we feel we may be under attack. This cognitive bias will help protect our self-esteem. If we attribute positive results to our individual characteristics, we give ourselves a pat on the back and a confidence boost.

By attributing failure to outside conditions or forces, we protect our self-esteem and clear ourselves of personal responsibility. This tendency to attribute blame for negative events to other people should come as no surprise, as few people want to be associated with negativity or be perceived as unsuccessful. People more often seek reassurance about how clever they are and certainly would be unhappy if people thought the opposite. Hence, rather than expose any possible negative traits and accepting that they did not pass the examination because, they are not as smart as they thought, a person will tend to try and shift blame by finding scapegoats for their lack of success. In contrast, people will take credit for positive outcomes, whether they are responsible for these or not, in the hope that this will reinforce their positive traits.

The problem with this type of behaviour is that, while it may protect a person's self-esteem in the short term, continuing to blame external factors means not accepting the opportunity for self-improvement. Choosing to blame others can lead to a person feeling powerless in addressing their problems. It is also clear that if somebody tries to continually take credit for things they have not done, there will be credibility issues sooner or later.

There is a positive side to self-serving bias inasmuch as the defence mechanism may help to motivate the individual in the short term. For example, if a person believes that a stagnant job market is preventing them from landing a good job, this may make them more persistent in their job search, at least for a while.

Good leaders will, therefore, look at themselves first if something goes wrong. They will want to fully explore the causes of the event; many times when things go wrong, this can be traced back to a management decision.

Activity

One clear implication of self-serving bias for the health and safety leader is that, when an incident occurs, people may try to protect themselves by blaming others or other external factors in the work environment.

What are the main negative outcomes from this? How do you think a leader might effectively deal with this challenge?

2.2 Decision-making processes, mental shortcuts, biases and habits

Hindsight bias

'With the benefit of hindsight, I wouldn't have done it that way.'

> Hindsight bias occurs when people feel that they 'knew it all along', that is, when they believe that an event is more predictable after it becomes known, than it was before it became known.[36]

Most people will be aware of situations where something unexpected or random intervenes that completely changes the outcome of an event.

In their research paper on hindsight bias, Roese and Vohs identify three inputs:[37]
- cognitive;
- metacognitive; and
- motivational.

The first of these inputs (cognitive) relates to the person selectively recalling information from past incidents; the information recalled is what the person now knows to be true (after the fact). The person will then try to make sense of this but will confuse it with what they thought they knew earlier.

Metacognitive inputs are defined as how easy a past outcome is understood and may be incorrectly applied to its assumed likelihood.

Motivational inputs are a result of people needing to see the world as orderly and predictable and not wanting to be blamed for anything.

When these three inputs combine, the result is that people tend to concentrate on a single point and neglect to take account of other reasonable explanations, therefore, applying hindsight bias.

Hindsight bias has three levels:
- memory distortion (I said it would happen);
- inevitability (it had to happen); and
- foreseeability (I knew it would happen).

Memory distortion relates to a person incorrectly recollecting an earlier personal judgement.

Inevitability involves personal beliefs about the condition of the world and that past events were predestined. Usually, the inevitable level also includes memory distortion as well as acceptance of beliefs about the factors that cause an event and which make certain outcomes seem more predictable than others.

Foreseeability is very subjective and come from beliefs about the person's own ability and knowledge. It involves the person believing that they could have foreseen an actual event. The foreseeability level also includes the inevitability level; it includes beliefs about the condition of the world as well as including belief about the person's ability to understand the world.

Ultimately, hindsight bias matters because it gets in the way of learning from our experiences. When discussing their research paper, Roese observed that:[38]

> If you feel like you knew it all along, it means you won't stop to examine why something really happened. It's often hard to convince seasoned decision makers that they might fall prey to hindsight bias.

Good leaders should not jump to conclusions when something goes wrong. A full investigation should be carried out, looking at the obscure facts as well as the obvious. This will allow learning from incidents and accidents.

2.2 Decision-making processes, mental shortcuts, biases and habits

Habits and decision making

When you think about a habit, what comes to mind?

One definition is: something that you do often and regularly, sometimes without knowing that you are doing it.

Habits are learned actions that have been reinforced in the past by a reward. They are triggered automatically when we encounter the situation in which we have repeatedly performed those actions before. They are largely outside of our conscious awareness and control. Habits are mentally efficient and allow us to conserve mental resource to use it for more difficult or new tasks.

Habits are likely to persist over time because they are automatic and do not need conscious thought, memory or willpower. Changing a habit takes time and requires conscious self-directed effort and planning.

Everyone develops both good and bad habits over time, but how does this influence decision-making?

If you think about some of the habits you may have, it will probably become clear quite quickly that these are, again, shortcuts to perform routine tasks without needing to concentrate too deeply on what you are doing. Habits are another way of making people more efficient. By turning routines and repeated behaviour into a habit, a person does not need to think much about making a coffee, what to have for breakfast or travelling to work. The brain acts automatically, allowing the person to concentrate on more important matters.

Habits are good because they mean that people do not have to dwell on or consider every single task in their day. However, because habits are part of the subconscious, discontinuing a habit, especially a long-term one, is not easy.

Developing good habits can help with the decision-making process. It allows people to focus on the bigger things that need attention and evaluation and stops them from getting caught up in routine activities that should not require a lot of concentration.

It is important for leaders to recognise that much of their behaviour is habit and that they will, therefore, be unaware of what they are doing. This may include taking risky or unhealthy shortcuts such as not using a handrail on steep stairs or poor habits when using display screen equipment. These bad habits could influence the workforce.

Activity

How do you think habits can affect the health and safety decision-making process?

2.2 Decision-making processes, mental shortcuts, biases and habits

Personal beliefs and how these can affect decision making

An individual's personal belief may or may not be the same as the belief held by the majority. Because it is personal it is based on the individual's perspective and formed as a result of the individual's experiences. This is important because belief is connected to values, attitudes and behaviour – factors which will impact decision making.

Individuals all develop personal beliefs over time. We may learn to believe certain things through life experience, we may be taught to believe things or may develop beliefs based on those held by our parents.

This individualised belief system can have a powerful influence on a person's decision-making. Therefore, health and safety leaders need to recognise the types of personal beliefs people have and how these might impact on decisions in the health and safety environment.

Familiarity

It is often more comfortable for people to deal with the familiar than risk the uncertain, or at least it seems that way at the time. Familiarity can often influence us to make assumptions that can be dangerous and misleading. Because a situation may seem familiar as it unfolds, a person may ignore the risks and jump to conclusion that they know what will happen next; they probably do not, but their beliefs tell them that they do. It is also known that humans underestimate risks.

Familiarity may also influence a person to stick with what they know rather than look at other alternatives. In an evolving and competitive environment, this may not always be useful.

Control acceptance

Organisations put controls in place to help protect their workforce. It is more likely that the workforce will use these controls if they believe that they will work. It is crucial that health and safety leaders know what these controls are and how to use them when required. This will have a positive impact on the workforce. Leaders must also ensure that all workers know what controls are in place, why they are there and, most importantly, how to use these safely and effectively.

Self-efficacy

Put simply, self-efficacy is about self-belief, which is a powerful factor in many decision-making processes. In order to overcome obstacles or challenges and achieve a desired outcome, a person needs to believe in themselves, to have confidence in their ability to set and achieve objectives. While, at some stage, this may be based on measuring previous experience, initially a leader must believe that they are capable of delivering a target. This belief is a very strong and positive personal belief and, if a person has a weak sense of self-efficacy, this may prevent them from even attempting to succeed in a task. It is important for a health and safety leader to recognise self-efficacy as a factor in the way people perform.

Responsibility

Responsibility is a key personal belief; this is especially so in effective health and safety leadership. A clearly-defined understanding that everyone has responsibility for health and safety will be an essential component in developing a successful health and safety culture. As a health and safety leader you should ensure that your teams believe they have the responsibility for health and safety. To do this you should be ensuring that they can take ownership of situations and make non-critical health and safety decisions for themselves. You also empower them to challenge others who are acting in an unsafe manner.

Element 2 Human failure and decision making

2.2 Decision-making processes, mental shortcuts, biases and habits

Normative beliefs

The statements 'what would they expect me to do?' and 'how would my team expect me to behave in this situation?' are examples of how normative beliefs work. They are based on the prevailing culture and how a person thinks others (probably others they care about or respect) would expect them to behave. Normative beliefs are important in a decision-making process because most people want others to approve of their decision. Everyone seeks approval of one sort or another and anticipating that our actions may affect the decision we make.

Consequences

The potential consequences of an action will affect decision-making; this is true for both uncomplicated choices and more complex processes. When evaluating potential consequences, people tend to consider not just what these consequences are but also when they will occur. There are occasions where unsafe behaviours will not have a consequence for some time, but others where the effect will be immediate. For example, an individual may believe that exposure to dust from not wearing respiratory protective equipment (RPE) may possibly lead to an occupational illness in many years, whereas not wearing RPE will definitely have the immediate consequence of getting the job done quicker. The consequences that a person believes to be immediate and certain are the ones that are most influential on their behaviour. There can be a tendency to think 'it won't happen to me' or that 'it won't happen immediately so I can leave someone else to sort it out'.

A health and safety leader should, therefore, ensure that safe and healthy behaviour is rewarded, recognised and praised at all times. This will help to ensure that the organisation is seen as a caring employer. Conversely, if workers know that a leader will walk past and ignore unsafe behaviours, this reinforces unsafe behaviours and can negatively impact on the safety culture.

Element 2.2 References

32 Kahneman, D, *Thinking, Fast and Slow* (2011, Penguin Books)
33 Chabris, C and Simons, D, *The Invisible Gorilla* (2010, Harmony) (http://www.theinvisiblegorilla.com/index.html)
34 *Oxford Advanced Learner's Dictionary* (https://www.oxfordlearnersdictionaries.com/definition/english/bias_1?q=bias)
35 Gruman, JA, Schneider, FW and Coutts, LM (Eds), *Applied Social Psychology: Understanding and Addressing Social and Practical Problems* (2017, SAGE Publications, Inc) (https://doi.org/10.4135/9781071800591)
36 Roese, N J, and Vohs, KD, "Hindsight Bias" (2012) *Perspectives on Psychological Science* 7(5), 411-426 (https://doi.org/10.1177/1745691612454303)
37 See note 36.
38 American Association for Psychological Science, '"I Knew It all Along ... Didn't I?" – Understanding Hindsight Bias' (2012) (https://www.psychologicalscience.org/news/releases/i-knew-it-all-along-didnt-i-understanding-hindsight-bias.html)

Element 3
Leadership

This Element will explore the different types of leadership styles. It will then examine the HSE's health and safety leadership model's five leadership values and supporting foundations. The five leadership values are the basis for the qualification assessment. Finally, the Element will look at how effective leadership communication can help to build relationships and rapport with the workforce.

The HSE's model is based upon an amalgamation of different types of leadership styles. The HSE model is an evidence-based model that draws upon HSE's research (including *Research Report 952*) which identified the key characteristics of good health and safety leadership.

Learning outcomes
- Demonstrate a range of appropriate leadership styles.
- Integrate the five values and supporting foundations of the HSE's health and safety leadership model within their professional practice.
- Practice effective leadership communication to build relationships with the workforce.

3.1 Leadership styles

There are many differing styles of leader and many theories of leadership, far too many to discuss here. However, it is important for any aspiring health and safety leader to appreciate at least a small selection of leadership styles, what they comprise of, and how they may be used to influence different groups of people.

When evaluating the impact of a leader, people sometimes wrongly believe that they have only one style. While this is possible, it is unlikely for an effective leader. One of the key skills in leadership is recognising how people are motivated in different situations and adapting the leadership style to fit the situation. It would be challenging and ultimately problematic for a health and safety leader to adopt a 'one size fits all approach'. Four different leadership styles will be considered here:
- Transformational
- Authentic
- Resonant
- Transactional

The four approaches can be compared and each encourages people to follow in different ways.

Transformational leadership

The transformational leader focuses on positively influencing people within an organisation to help and support each other as well as the organisation as a whole. For this to take place the transformational leader must build respect and loyalty in those who follow. By creating a culture based on admiration and trust, this theory suggests that people will be willing to work harder, support one another and give more to the organisation than would otherwise be the case.

In short, the transformational leader will generate trust, respect and admiration from followers which are considered important facilitators to motivate people to perform beyond expectations. They can have a positive impact on safety by leading by example and acting as safety role models; they will demonstrate a high priority for safety over other organisational goals. In addition, they encourage workers to work toward higher standards of safety and to try out new ways of working safely. They demonstrate a real concern for the well-being and safety of workers and will demonstrate respect for the views and opinions of others. They have a positive influence on safety by enhancing perceived fairness and worker organisational commitment and creating a positive safety culture. The transformational leader will be a supportive presence and is unlikely to be egotistical or self-important as both these attributes will have a negative impact on transformational leadership.[39]

3.1 Leadership styles

Here are some of the characteristics of transformational leaders:
- organised and inspire creativeness from followers/colleagues;
- works well with teams, workers will identify with them and they encourage followers to work together to achieve the best results;
- respectful to others and therefore respected by others;
- they will articulate a vision that followers can aspire to and seek to attain;
- they challenge assumptions and traditional ways of doing things, invite new ideas and encourage followers to 'think outside of the box';
- create a supportive climate and promote learning opportunities to meet the followers' needs;
- takes responsibility for actions as well as encouraging others to take responsibility; and
- creates a culture of respect based on rapport and positive relationships.

Activity

Why do you think this style of leadership may be useful for the health and safety leader?
Can you think of any negative aspects to this method?

3.1 Leadership styles

Transactional leadership

Transactional leadership will, for many people, probably appear familiar and is what conventional leadership has traditionally looked like in many organisations. The transactional leader is all about the status quo and the tools used will include:

- **contingent reward** – leaders agree with followers' specific goals with commensurate rewards for effort and commitment;
- **active management** – leaders actively monitor the workforce to ensure their behaviours, routines and processes comply with expected standards and intervene before problems arise (leaders check that the rules are being followed); and
- **passive management** – the leader will usually only intervene after a problem has occurred.

The transactional leader is more likely to be found in established organisations, where the workforce is used to patterns and tried-and-tested methods of improving productivity. The motivation here is almost entirely to do with self-interest or the fear of consequences of under-performance or non-compliance.

Transactional and transformational leadership theories are often compared, and the following table shows the main differences.

Transactional leadership	Transformational leadership
Leadership is responsive.	A proactive leadership process built on rapport.
Works well within the organisational culture and hierarchical structure.	New ideas from various sources are implemented to bring about change to the organisational culture.
Workers accomplish objectives through either rewards orpenalties set by the leader.	Appeals to the workforce's moral values or higher ideals to achieve organisational and individual objectives.
Motivation is based on rewards designed to appeal to worker's self-interest.	Motivation is based on the interest of the group or the bigger picture rather than individual self-interest.

Activity

Much of the writing around motivation theory and leadership styles places more emphasis on the value of transformational or authentic styles. This suggests that real motivation factors are more to do with self-development than material wins.

Do you think this is valid?

Does the transactional method have value, if so in what circumstances do you think it is most appropriate?

How might you adopt this style as part of your health and safety leadership communications mix?

3.1 Leadership styles

Authentic leadership

Authentic leadership is a relatively new leadership style built on extremely old theory or principles; these pre-date many of the more conventional theories by many centuries. The roots of authentic leadership are based on ancient Greek philosophy 'know thyself'. Authentic leaders are defined as those who are self-aware, confident, genuine, optimistic, moral/ethical, balanced in terms of decision-making, and transparent in enacting leadership. Research shows links between authentic leadership and positive safety climate.[40]

There are obvious similarities between authentic leadership and transformational leadership. Certainly, the key attributes of the authentic leader would be a great advantage in transformational leadership.

These include:
- Self-awareness ('know thyself'). This is a prerequisite for being an authentic leader. This is about understanding your own strengths and weaknesses as well as recognising your values. Authentic leadership requires that you really know what you stand for, alongside honestly recognising what you value is critical. Becoming self-aware is the first step in developing the other components that make up authentic leadership.
- Relational transparency or being genuine. This is about being genuine and honest in your relationships with others. A straightforward approach with no concealed agendas or power games. The authentic leader will ensure that people will know where they stand whether they like it or not. Honesty sometimes means making tough decisions.
- Balanced processing or fair-mindedness. This means listening to others, including opposing viewpoints that are actively sought by an effective authentic leader. Planning is a key characteristic with available options being discussed before choosing a direction. This is not a solitary enterprise and everyone's views are important.
- Internalised moral perspective or doing the right thing. Ethics and fairness are driving forces of the authentic leader, with no room for personal gain at other people's expense, dishonesty or exploitation. Leadership here will have goals that are not self-serving and will depend on the authentic leader's moral compass to ensure the right outcome for all concerned.

The authentic leader will set ambitious standards for themselves as well as everyone else. Success will be a team effort and based on the principle of win-win, something that is not always easy to achieve in many organisations. However, this style of leadership has gained ground during the past 30 years in many areas – following an authentic leader can be a rewarding and satisfying experience.

It is challenging to attempt to be an authentic leader, it takes strength of character to recognise your own shortcomings. You need to really reflect and get to know yourself and develop the courage and strength to do the right thing. In truth there are many leaders who are far from authentic, and clearly corruption and lack of morality are all too common, hence the search for good people to lead is becoming compelling.

Activity

How do you feel about the idea of authentic leadership in health and safety?

How important do you think it is that a leader demonstrates their intention to do the 'right' thing?

What other benefits might this bring to an organisation as well as individuals?

3.1 Leadership styles

Resonant leadership

During the past 20 years there have been some very substantial changes in leadership and management styles as well as organisational and cultural behaviour. This has included a significant and growing awareness of the usefulness of tools and techniques in business and the world of work. Resonant leadership has three dimensions:
- mindfulness;
- hope; and
- compassion.

Mindfulness – leading a life by developing a complete and conscious awareness of self, others, environment and work; being awake and being aware.

Hope – enables us to believe that our goals are achievable; this motivates us to inspire others to reach those goals while striving to achieve them ourselves. Others may dream about a better future and believe they can attain this and will, therefore, develop an optimistic point of view.

Compassion – helps us to include emotions when thinking, making decisions and taking action. The resonant leader will empathise with others and will try to put themselves in another's position; they will treat superiors and subordinates equally with empathy and compassion.

Resonant leaders will be in tune with those around them and the leadership style is based on the concept of emotional intelligence that has been championed by, amongst others, the US scientist and psychologist Daniel Goleman.

Emotional intelligence (EQ or EI) is a term created by two researchers – Peter Salavoy and John Mayer – and popularised by Dan Goleman in his 1996 book of the same name.[41] EI can be defined as the ability to recognise, understand and manage our own emotions and to recognise, understand and influence the emotions of others.[42]

In practical terms, this means being aware that emotions can drive our behaviour and impact people (positively and negatively) and learning how to manage those emotions – both our own and others – especially when we are under pressure.

There are obvious parallels with some of the concepts discussed earlier in the sections on transformational and authentic leadership as emotional intelligence requires that a leader understands and has control over their own feelings and drives to effectively understand how to best have an impact on the behaviour of those around them.

Emotional intelligence is also a key ingredient in resonant leadership which Goleman suggests comprises four styles as shown in the following diagram.

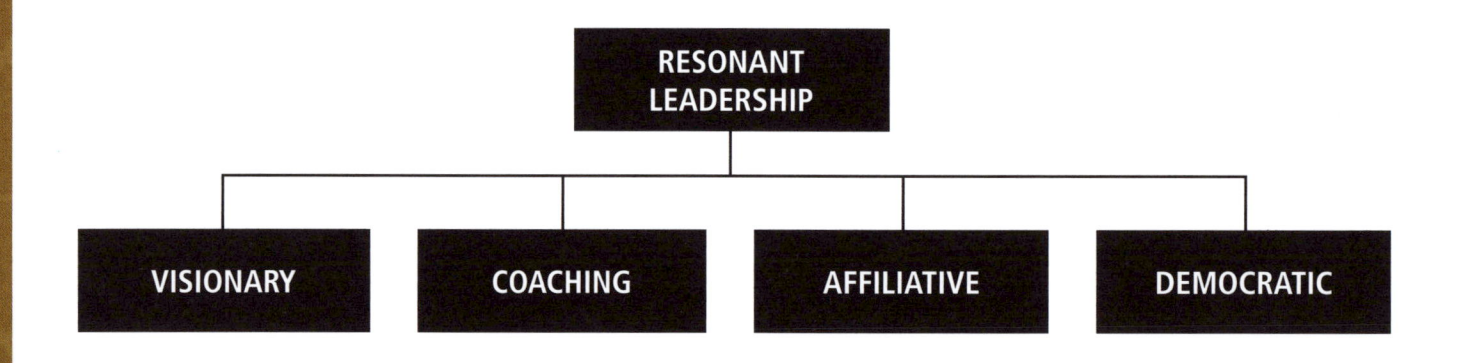

```
                    RESONANT
                   LEADERSHIP
        ┌──────────────┼──────────────┐
        │              │              │              │
   VISIONARY       COACHING      AFFILIATIVE     DEMOCRATIC
```

Element 3 Leadership

3.1 Leadership styles

Visionary

Visionary leaders know where they want go, share that vision and take people with them by inspiring them to want to follow. While the visionary leader will lead from the front, they will expect those who follow to think for themselves and understand the part they are playing in the bigger picture. This will build resonance by making people feel part of something and motivating them to achieve a shared goal. This style of leadership is suitable when organisational changes are taking place or when a clear direction needs to be communicated and implemented effectively.

Coaching

Coaching is a valuable tool that focuses on the individual personal development of those in a team. Using a coaching style of leadership will demonstrate genuine interest in the workforce and thus enable leaders to build trust and rapport. Coaching has been proven to motivate people to achieve more, building resonance by linking people's needs with organisation's goals. Coaching is highly effective in improving worker performance and conveying a sense of value to the individual. The following diagram shows some of the benefits of coaching.

COACHING

MOTIVATION • COACH • POTENTIAL • DEVELOPMENT • SKILL • SUPPORT • KNOWLEDGE • ADVICE

Affiliative

An affiliative leader will strive to build collaborative relationships using empathy. It is key that the leader demonstrates that they really value others and cares about their feelings. By linking people together, the affiliative leader builds resonance by creating harmonious relationships that promote interdependence. This style of leadership can help with bringing together or strengthening a team and building performance over time.

Democratic

A democratic leader will appreciate the importance of understanding the feelings and valuing the opinions within a diverse group. They will use the expertise and knowledge that exists within the group to achieve shared goals through collaboration and teamwork. The democratic leader will need to be a highly effective communicator with particularly good listening skills. Empathy is a key attribute for the democratic leader.

By valuing people's input and attaining commitment, the democratic leader can build resonance. They can obtain buy-in from team members for projects or or organisational change.

3.1 Leadership styles

Activity

What do you think are the key benefits for adopting a resonant or transactional leadership style in health and safety?

Can you think of any examples of where this style of leadership may be essential to getting the job done?

Can you also think about why it may need to be carefully managed?

Case Study

A construction company decided that they needed to get the workforce more involved in decision making.

In the past, the management of the organisation was resistant to the idea of worker involvement in policy making and suspicious of any perceived dilution of its authority.

The company changed its Board and took this opportunity to involve the workforce more in consultation.

> The new Board of Directors understood what we wanted to do and were happy to support the idea of worker engagement/consultation. This meant that we had a 'blank canvas' on which to place whatever engagement/consultation mechanism we felt was most appropriate, so obviously it was important to get this right.

The company formed a Consultation Committee which, other than the Health and Safety Director, was formed entirely from the workforce. The company has benefited from this change; workers are owning health and safety and have seen improved relationships throughout the organisation.[43]

Element 3.1 References

39 HSE, *A review of the literature on effective leadership behaviours for safety* (RR952, 2012) (https://www.hse.gov.uk/Research/rrhtm/rr952.htm)

40 Birkeland Nielsen, M, Eid, J, Mearns, K and Larsson, G, "Authentic leadership and its relationship with risk perception and safety climate" (2013) *Leadership & Organization Development Journal*, 34(4), 308-325 (https://doi.org/10.1108/LODJ-07-2011-0065) and Borgersen, HC, Hystad, SW, Larsson, G, & Eid, J, 'Authentic Leadership and Safety Climate Among Seafarers' (2014) *Journal of Leadership & Organizational Studies*, 21(4), 394-402 (https://doi.org/10.1177/1548051813499612)

41 Goleman, D, *Emotional Intelligence* (1996, Bantam Books)

42 Institute for Health and Human Potential, 'What is Emotional Intelligence?' (https://www.ihhp.com/meaning-of-emotional-intelligence/)

43 HSE, 'Case study: Bardsley Construction Limited' (https://www.hse.gov.uk/involvement/assets/docs/bardsley.pdf)

3.2 The five leadership values and supporting foundations

The HSE's five leadership values were set out in the Introduction. To recap, these are:
- building and promoting a shared health and safety vision;
- being considerate and responsive;
- providing support and recognition;
- promoting fairness and trust in relationships with others; and
- encouraging improvement, innovation and learning.

Underpinning these values are three foundations of leadership. These are:
- involvement and communication;
- effective role modelling; and
- embedding robust health and safety management as a business norm.

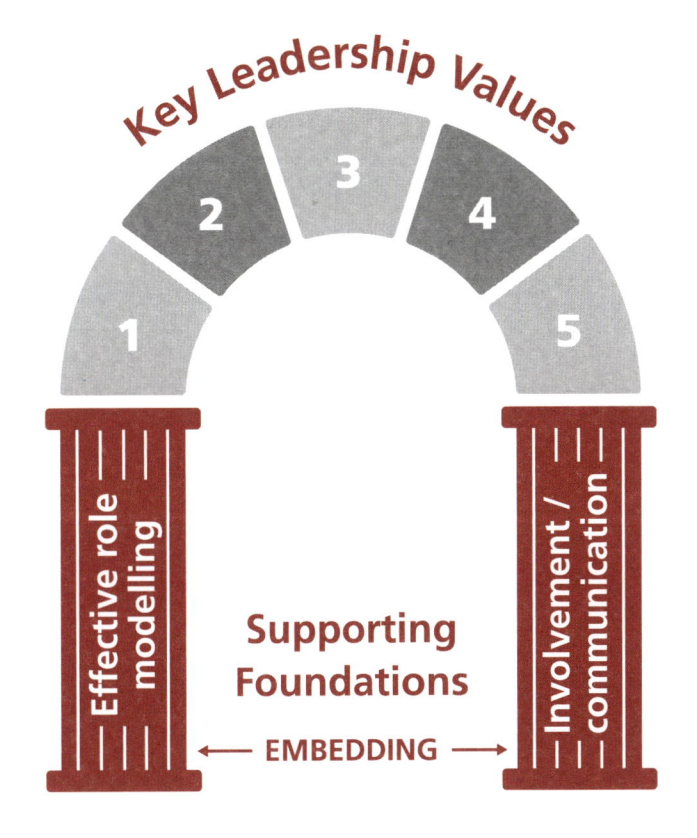

Activity

Think about your own role within your own organisation in relation to the leadership values and foundations.

Do you think that all of these principles are applied?

Can you provide some examples of this? If not, what might be the reasons for this?

3.2 The five leadership values and supporting foundations

The health and safety leader must be proactive across the whole organisation. The leader must ensure that their organisation is not just talking about the subject but that health and safety is incorporated into the business. The health and safety leader should be passionate and vocal about health and safety to ensure that it stays at the core of the business. The leadership values will ensure that every part of the workforce can play a part in the delivery of the highest standards of health and safety.

Involvement and communication

As a health and safety leader, you should be proactive in creating and using opportunities to have two-way conversations with all workers relating to health and safety issues. Health and safety messages should be consistent and targeted and tailored to the intended audience within the organisation. It is also vitally important that you make sure that the messages contained within any communications have been received and understood.

It is also important to involve the workforce in all aspects of health and safety where and when relevant and appropriate to do so. Empowering the workforce to make decisions is very powerful. Not only will this make individual workers look at their own health and safety behaviours it will also make them look at their colleagues' behaviours as well. This will give your workers the confidence to challenge those who are acting in an unsafe manner.

Effective role modelling

As a health and safety leader your behaviour will be witnessed and judged by the rest of the workforce; it is, therefore, very important that you lead by example, for example, by wearing the correct personal protective equipment. It is very unlikely that the workforce will follow safety rules if they see that you are breaking them consistently, or if you ignore others who are carrying out unsafe actions. You should take every opportunity to demonstrate your personal commitment; the more you do, the more it will be noticed. Having leaders visibly demonstrate safe behaviours shows the workforce that it is not one rule for them and one rule for the organisation's management.

Embedding

Embedding is about ensuring that health and safety is not considered as a stand-alone activity. It is imperative that health and safety is embedded into the organisation at all levels and that it is considered critical for business success. Health and safety should be seen as just one tool to help support your organisation's purpose.

3.2 The five leadership values and supporting foundations

Being considerate and responsive (health and safety leadership value 2)

People generally respond well to people who demonstrate empathy towards them. Certain leadership styles and theories are built on developing skills of empathy and emotional intelligence.

There are many methods of demonstrating responsiveness and consideration; the goal should be treating people how you want to be treated yourself. Listening and responding appropriately are skills a good leader should possess. This of course, does not mean that as a leader you should agree with everything everyone says, but it does mean you should listen and demonstrate that you are interested in everyone's views or concerns. Equally, responding to other people's needs will continue the good work of building effective relationships. If a training need is identified, within reason this need should be met. If there is a gap in the knowledge or skill set of people in the workforce that prevents them from undertaking the tasks required of them as part of the health and safety vision, training or coaching should be made available where and when possible. This is about investing in the people you need to make the vision happen.

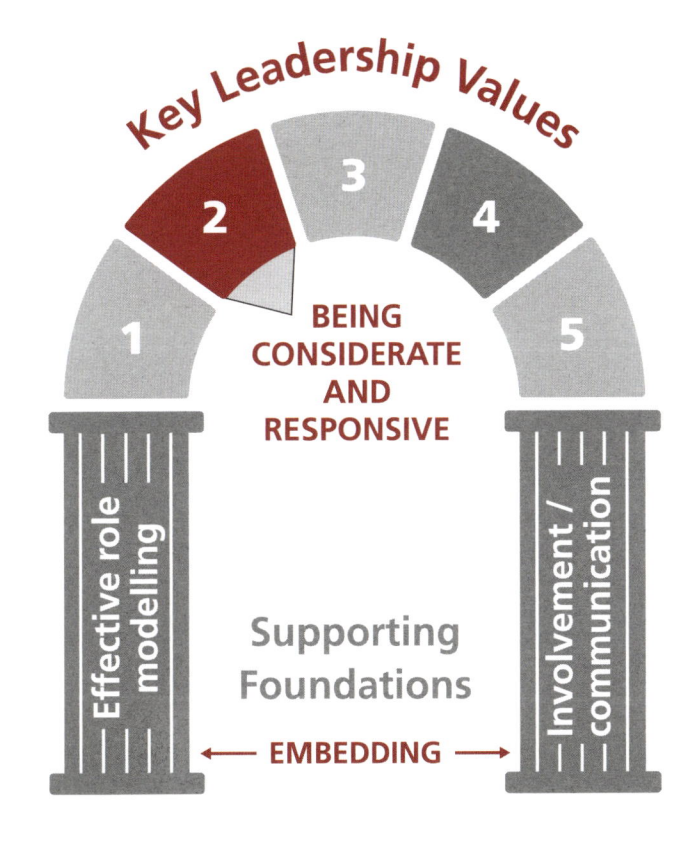

Assessment Activity
- Please refer to the document Unit HSL1, guidance and information for candidates and internal assessors.

You should now complete task L2: Being considerate and responsive.

Element 3 Leadership

3.2 The five leadership values and supporting foundations

Assessing own health and safety leadership performance

One of the most important things that a health and safety leader should do is to critically evaluate themselves. Our perception of ourselves may be very different to the way others in the workforce perceive us. While most people will have some form of upwards performance appraisal, not everyone will receive an evaluation from the rest of the workforce (often called a 360 review).

There are two important things that you can do to improve your leadership style. Firstly, question the workforce. A good frequency for this is every six months. It is important that the workforce know that there will be no recriminations from their responses. You should ask the workforce the right questions; open questions usually produce a better response than a closed question. So, for example, don't ask the workforce "am I a good leader?"; inevitably the response will be 'yes' and this does not help anyone. A much better question would be "What one thing do you think I could do to make me a better leader?" However, it is not just about asking the questions, it is also about listening to the response responses, considering the feedback and addressing any criticism.

Secondly, practice self-reflection. The process involves considering any actions you have taken during a period (perhaps every six months) and analysing them to find out whether you could have done anything better or differently. Very often, carrying out some form of continual professional development will help with this exercise.

"Who would like to give me feedback on my leadership qualities?"

As a leader you should be striving to always improve your performance and how others see you. Undertaking these activities will help you to achieve this goal.

Element 3 Leadership

3.3 Building relationships with the workforce

Creating and sustaining a positive relationship with the workforce is a key consideration for leaders within most organisations. An effective health and safety leader will understand that maintaining a dialogue with the workforce provides the right environment for sharing and promoting the health and safety vision, as well as ensuring that effective communication is taking place.

There are different ways in which effective relationships can be formed and maintained using a variety of tried and tested leadership techniques and strategies.

Leadership walkabouts and rapport

One very successful strategy that may be employed is managing by walking about (MBWA) (or wandering around). MBWA is an accepted management technique. This has not always been the case; traditionally organisations have had strict hierarchical structures and much more of a closed door policy than is normal today.

The idea of managers walking around in the working environment engaging with the workforce is now considered the norm. This creates an environment where workers feel comfortable raising concerns. Maintaining an organisational culture that encourages managers to engage directly, frequently and informally with front line workers, asking for their opinions and encouraging them to raise any issues, is very well suited to the MBWA technique.

Walkabouts should not be treated as a tick box exercise to meet management KPIs. They need to involve two way communication and should also consider the value of the walkabout from the workers' point of view.

Walkabouts also provide an excellent learning opportunity for managers to observe workforce activity first hand and work directly with front line workers to resolve problems.

The benefits of using this technique to establish rapport with people from right across an organisation should be self-evident, after all it is widely accepted that face-to-face communication is often the most effective way of delivering and ensuring information has been received and understood. So, what is rapport and why is it important in health and safety leadership?

Key Term
Rapport

A close and harmonious relationship in which the people or groups concerned understand each other's feelings or ideas and communicate well.[44]

By establishing rapport with colleagues and the workforce, the effective health and safety leader is able to understand other people's feelings and communicate well. These are both vital factors for the successful management of health and safety.

It is important to remember that while MBWA should be frequent and largely informal, focus is also important. It is also worth remembering that most effective communication is planned.

In broad terms, planning for leadership walkabouts can be divided into four sections:
- frequency
- format
- lessons learned
- taking action.

3.3 Building relationships with the workforce

MBWA frequency

Being effective and building positive relationships requires more than just walking around randomly talking to a few people every now and then. Rapport is built on reliability, trust, active listening and continuity, through regular, positive encounters with the workforce.

The appropriate timing and frequency of these conversations will depend upon the needs of the organisation and the individuals involved. They should not become a barrier or an unwelcome interruption to the flow of work. However, if people are to be encouraged to interact and share thoughts, a good strategy is to have a regular pattern for the conversations. This could be a specific time during the working week when certain teams may be more responsive or less busy.

MBWA format

Once timings and frequency have been established, the next consideration is how the activity will take place. This will depend on the circumstances of the organisation, its environment and the industry in which it operates. Establishing a format that people can identify with and can be used to best effect will be essential if you are to establish a good rapport and make the MBWA activity worthwhile.

For example, the format could be based on the 'Safe Deal'[45] cards that have been produced by the Health and Safety Laboratory. These cards contain questions that can be used to open a dialogue with the workforce which can be either formal or informal. Alternatively, you could devise your own list of questions to use as a prompt when you are walking around, or you may wish to use a tick list. Open questions using the TEDS (Tell, Explain, Describe, Show) framework work best. Some questions you may want to consider are:
- 'Are we looking after your health and safety?'
- 'What would make being healthy and safe easier?'
- 'What would you like to change about health and safety here?'
- Ask in the third person: 'why would someone do that job like that?'
- 'What's slow, inconvenient or uncomfortable about doing this safely?'
- Ask 'is there anything you do where you tell yourself "it will never happen to me"' and, if so, ask 'what if it did happen?'

Ultimately, the walkabout should produce tangible outputs. It is a continuous process that will involve giving and receiving information, acting on this information, reflecting on what happens and understanding what should happen next; a continuous process.

Devising the format of your MBWA is an essential step because any rapport or positive relationship that is developed by asking questions and getting answers will be severely damaged by not taking action or providing feedback.

Activity

What does leadership walkabout mean to you?

How do you apply it/what format do you use?

What more could you do?

When do you do your walkabouts?

3.3 Building relationships with the workforce

Lessons learned

A key part of MBWA revolves around gathering information. An effective health and safety leader needs to develop a better understanding of the workforce's attitudes or behaviours towards a particular process and to learn how particular activities are prioritised. Having clear and accurate records of the interactions that took place during an MBWA activity is an effective way of encouraging people to participate in the decision-making process. This will allow all those involved to learn from the process and feed this information back into the cycle.

Lessons learned will be of significant value in making future decisions or planning how to introduce new health and safety initiatives. The reflective process will add value to the technique and, used in conjunction with frequency and format, will provide valuable lessons and modifications for future practice.

Taking action

One valuable method for taking action is by conducting a SWOT analysis of a process, product or service. SWOT is a simple model that looks at Strengths, Weaknesses, Opportunities and Threats.

However, if there are no tangible outputs from this exercise, there is little value in spending time carrying it out. If there are no outputs that the workforce can see, they are less likely to engage with you in the future.

The same is also true about any ideas for improvement received from the workforce. Even if the idea or suggestion has been considered previously, it is always important to feedback to the workforce and to let them know whether the idea will be taken forward or, if not, why not.

STRENGTHS
- Advantages
- Capabilities
- Resources

WEAKNESS
- Disadvantages
- Vulnerabilities
- Limitations

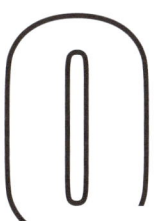

OPPORTUNITIES
- Chances
- Developments
- Benefits

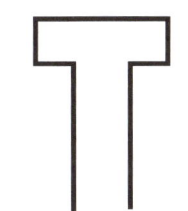

THREATS
- Obstacles
- External effects
- Risks

Activity

How might you use the SWOT tool to analyse the MBWA method?

What might the strengths be of such a process?

Where might it fall short (weaknesses)?

What type of opportunities could be revealed?

Which areas might produce a threat?

Element 3 Leadership

3.3 Building relationships with the workforce

Effective communication in building rapport

Rapport is primarily about building trust and respect to promote good two-way communication.

Building rapport will help ensure that the messages delivered by the health and safety leader are:
- received;
- understood; and
- acted on.

The same is also true for the messages delivered by the workforce to the leader.

This requires that effective communication takes place. But, what is effective communication and how do you ensure you are doing it right? Sometimes it is easy to think that because you have said or asked something, everyone has understood your message. This is not true.

Effective communication is far from simple. There are many everyday factors that can complicate the communication process and prevent messages from being received in the way that was intended. These factors (or 'noise') can result in misunderstandings, have a negative effect on relationships and make it difficult to build a good rapport. Some of these factors include:
- time (as in timely delivery as well as lack of time)
- place (environment)
- physical noise
- language (both spoken and body)
- attitude
- stereotyping or prejudice
- stress or pressure of work
- culture.

Activity

Make a list of some of the barriers to effective communication (either face-to-face or other means such as email) you can think of. These might be physical and may include things like attitude, environment, culture, stereotyping.

3.3 Building relationships with the workforce

Barriers to building a good rapport with the workforce

Communication requires understanding and practice to get it right. Some people are natural communicators, but most people need to learn how to communicate effectively in the same way as they would learn to speak another language or any other skill.

There are always barriers to communication. Some of these occur on a daily basis and not all of these can be eliminated. However, it is possible to reduce the negative influence of these barriers on communication in the working environment by identifying and considering barriers and adapting the message and its delivery.

Many 'audiences' are not really audiences. Quite often, they are not ready or not waiting for your message. Some audiences may be ready for you but too busy. However, they may also be irritated, distracted or simply uninterested. A good leader will recognise this and plan for it.

Health and safety leaders need to plan communications to reach the whole audience; your message could lose impact if just one person does not engage or starts to look bored.

There are also different audiences to consider. Within each separate group, there will be a range of attitudes, beliefs and behaviours. Being aware of this when communicating may help you to interpret or understand the responses that you get. It is helpful to remember this during meetings, consultations and encounters with the workforce.

The attitude and behaviour of those with whom you need to communicate can also affect the impact of your communication. This could include:
- a history of negative attitudes or a culture of mistrust; and
- previous management failure to build rapport or trust with the workforce.

The physical environment can also affect the impact of the communication. If the environment does not work and there is nothing that could realistically be changed to improve it, it may not be a good idea to start the conversation. When reviewing the environment, consider:
- distractions, interruptions and noise;
- whether there is a lack of resources;
- occupational stress; and
- workspace, layout, comfort.

When using the MBWA technique you will need to be mindful of these issues.

Element 3 Leadership

3.3 Building relationships with the workforce

What good communication looks like

In planning communication and conveying information, a useful guide that will help ensure that the right information is given is the seven Cs of effective communication.

Clear	Leave no uncertainty, provide facts.
Concise	A brief explanation, no unnecessary information provided.
Concrete	Solid, dependable and real.
Correct	Is it right in presentation, delivery, style, and time?
Coherent	Does it make sense, really make sense?
Complete	Nothing missing, no chance of misunderstanding?
Courteous	Friendly, open and honest, no hidden agenda?

Applying these rules can help ensure that information is right and suitable for purpose, therefore, avoiding misinformation or difficulty interpreting what is needed.

3.3 Building relationships with the workforce

Physiology, voice, words

It is also important to consider how you look when you are speaking (physiology), how you sound (voice) and the words that you use. Physiology is most important, followed by voice and then words.

With physiology and voice it is important that you try to do this to put people at ease. However, you must try not to be too obvious about this as your audience may think that you are deliberately imitating them to make fun of them and they could become immediately hostile. Leaders should ultimately be trying to persuade the audience to take part in effective communication.

Physiology
When face-to-face with an audience it is important that you try to match and mirror your audiences':
- posture;
- gestures;
- facial expressions;
- eye contact; and
- breathing.

Voice
You should try to ensure that you match and mirror your audiences':
- tone of voice;
- pitch;
- frequency;
- speed;
- timbre;
- characteristics;
- quality; and
- volume.

3.3 Building relationships with the workforce

Words

For each audience:

- consider the content and what words you will use (for example, do not use street slang when addressing a board of directors);
- avoid using jargon unless you are sure that your audience knows and understands what you are talking about; and
- try to include any shared experiences or interests to put your audience at ease.

It is easy to sometimes forget the most important part of communication – the receiver or audience. If they do not let the speaker know they have received and understood the message, communication has not taken place. This is where the importance of rapport becomes even clearer. In order for a piece of communication to take place it requires a loop:

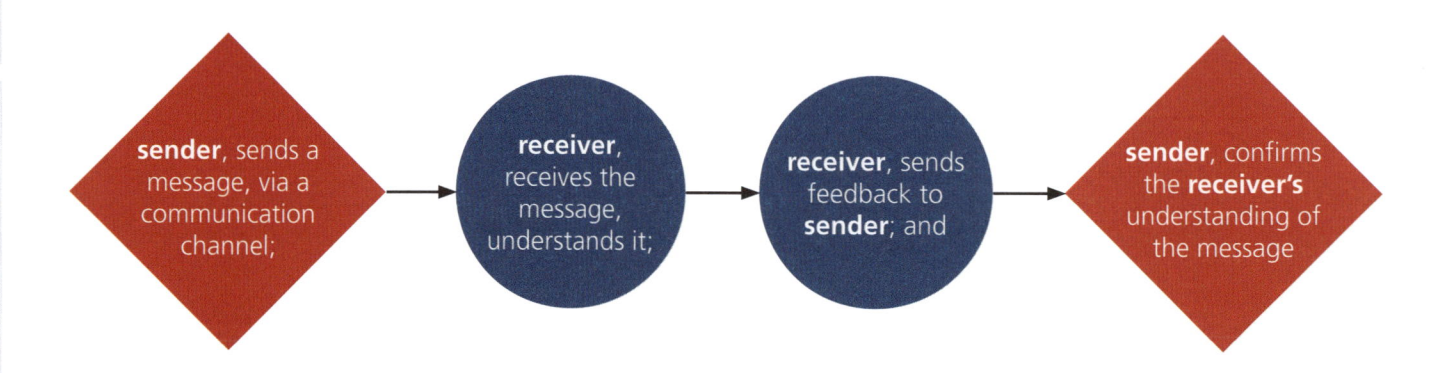

sender, sends a message, via a communication channel; → **receiver**, receives the message, understands it; → **receiver**, sends feedback to **sender**; and → **sender**, confirms the **receiver's** understanding of the message

This can be as simple as:

Sender:	'I will meet you in the coffee shop at 10.00'
Receiver:	'What, the one on the high street next to the station at 10.00 this morning?'
Sender:	'Yes'

This is a complete communication loop. The same feedback rule would apply to a complex email or detailed conversation. Unless there has been confirmation of understanding and a signal of action, communication has not taken place.

Face-to-face communication has the advantage that the sender and receiver(s) are occupying the same physical space at the same time, hence MBWA eliminates many of the barriers to communication. It is easier to gauge whether your information has been received correctly when your audience is in front of you. It is essential to get the receiver to confirm the message, especially where safety critical information is involved. This method can also be used during non-face-to-face verbal delivery such as a telephone call or video conference. You may want to ask for a written response to any written communications especially those that are sent by email, text message or similar.

3.3 Building relationships with the workforce

The right information, praise and questions

The point of any communication is to provide accurate information to another party (audience). It is, therefore, essential that you provide your audience with the right information the first time you send it. Your workforce is unlikely to have confidence in your leadership if, for example, you have to keep sending emails to correct previous emails that contained incorrect information.

Another part of communication is using it to praise people. If you see someone doing something to improve safety or challenging someone else's behaviour, praise them. It does not matter when you do this or what communication method you use, as long as it is done (receiving praise makes us feel better). In other cases you should be asking questions and generally showing an interest in the way the workforce are working. The more you ask, the more the workforce will remember you. This could have a positive impact going forward as workers will see you as someone who is open to communication and someone who will take account of their views.

When you want to tell someone something important, consider the timing of the message and ask yourself the following questions:

- How is time going to affect the medium you choose?
- Is this the only or best time to communicate?
- Can you put all other concerns aside for a minute?
- Can you help your audience to do the same?
- Is everybody clear about how long this will take?

It is not practical to expect to have a perfect time and place to deliver important information to the workforce or to engage in conversations with them. But it is worth considering the ways in which barriers can be minimised to obtain the best result possible. The successful health and safety leader will plan to mitigate as many negative effects as they can to ensure the maximum impact of information given.

Activity

Information overload: do you sometimes feel overwhelmed by the amount of information you are expected to process?

How much of it is of real value?

How do you prioritise what you need?

How do you know what you need, without examining everything?

3.3 Building relationships with the workforce

How information can be given

There are some crucial factors that can impact on how information that is communicated is received. It is important to remember that communication has only taken place when the message has been received and understood and that feedback has been received that this has happened..

There are certain things that should be taken into account when giving information to others. It is also important to understand how to effectively gather useful information.

Primacy/recency effect

If you consider the way you personally absorb information, it will provide you with some good ideas about how other people may respond to your own efforts to communicate information. Most people share characteristics in the way they learn and take things in. Take for example attendees at a short training course. At the start, most people will be alert and ready to receive the information, in the right state of mind, prepared and hopefully, interested. So, there will be focus on the information at the beginning and this will normally mean it will be remembered. This is the primacy effect. At the end of the course there will be summarising and discussion, again because this is fresh in the mind, people will generally pay attention here and the information will stick; this is called the recency effect.

What this means is the process of giving information is reflected in many processes, from learning theory to advertising. It is important to state the important things at the start of the communication process (primacy) and, at the end, recap the important points (recency). This should ensure that the essential information is received and committed to memory or acted upon.

Language and jargon

Language and jargon play a part in giving information. Language in communication terms is more than just choosing the right words to convey meaningful information. Language is also to do with dialect, perception, physical or body language, levels of understanding and the use of jargon. To give information effectively it is essential to know your audience – consider who they are, what they do, what they know, how they feel and how they will feel about the information to be communicated. Knowing your audience will help you to use the right language. It will allow you to judge how technical your message should be, to assess the correct tone for this particular group and very importantly, how much jargon you should use, if at all?

Jargon is either a useful shortcut for people with a shared understanding or knowledge of a subject, or a baffling and extremely irritating barrier to communication. This can simultaneously embarrass and alienate people as well as ensuring that any information you provide is either misunderstood or ignored. If you are to use jargon that includes acronyms and slang, you need to be absolutely certain that the people you are talking to understand what you are saying. If there is any doubt, leave it out!

Element 3 Leadership

Who delivers the message?

For information to be received and acted upon correctly, it should be delivered using a credible source, because the messenger delivering the information may be as significant as the information itself. This means ensuring the right person for the right audience. This may not be you, it may require that you find someone who you know is trusted by the particular audience you need to engage with. It may be that several people will need to be briefed and they in turn will convey the vital information to specific groups using the appropriate language and communication techniques. This could range from a senior manager to a local worker representative addressing a team, or it could be a health professional such as an occupational health nurse or a respected peer. It may require that some training is given to the messengers specific to the information being communicated. It is important that the information is tailored for each audience.

Making it memorable/fun

If you ever wonder why you remember some messages much more than others, irrespective of how important they might be, it could be something to do with the way the message was delivered. When communication is relevant, worthwhile and to the point, it becomes compelling and provokes a (positive) response. Just because a piece of information is important does not mean it cannot be delivered in an engaging or entertaining way. The task is to get people to remember what you need them to remember. How you do it should be based on the audience you are attempting to engage with, their attitudes and behaviour, the type of media they are used to, their attention span, the complexity of the information and how you want them to respond.

A useful device that can be used here is called PASS.

Purpose	What is the information meant to do?
Audience	Who are the audience?
Structure	How will the information be structured (what type of delivery method will be used)?
Style	What is the appropriate style for this audience (formal, informal, chatty or official)?

Presenting the same information in different ways

You may have to present the same information in different ways to ensure that each target audience gets the message the way it was intended. For example, people who regularly use social networks or visit YouTube to help solve problems or obtain information may not respond well to printed notes or a formal style report. Those people who read and respond to formal reports and specifications will be more comfortable with this style of communication.

Tailoring the information

Not only should you consider presenting information in different formats, but you should also think about tailoring the information to the audience. This will ensure that people will receive information that is relevant to them and that they understand the information. For example, there is no point sending an email to all workers (which might include functions like finance and packing sections) containing instructions on a new maintenance strategy that has been implemented. If these types of communication become commonplace, workers may start ignoring them as a matter of course because they will make the assumption that there is nothing relevant in it for them. This could, obviously, have a major impact for any industry but especially in the high hazard industries.

3.3 Building relationships with the workforce

How to gather information

A good health and safety leader must ensure that they have gathered the right information before making any conclusions or decisions about situations.

The right questions to ask about work activities and when to ask them

It is important to not only ask questions of your workforce, but also that the question is the right question. Open questions are usually better to ask than closed questions. Open questions require a response other than 'yes' or 'no' and, therefore, let the worker elaborate on their answers. However, there are times were mixed questioning is appropriate. Closed questions are appropriate when you are confirming a piece of information given to you for example, 'You are confirming that there were no reported faults when the equipment was started up?' to which the respondent would answer 'Yes'.

Good health and safety leaders will want to understand why things are done in a particular way. You should not be assuming that because it has always been done that way that this is the only way to work. So, for example, good questions to ask your workforce could be 'what other way could you carry out that activity that may make it safer for you?' or 'what other equipment would help you to carry out the activity more efficiently but safely at the same time?'. The best time to be asking these questions is during your leadership walkabouts. It may be advisable to have a list of questions with you to use as prompts with you when you are on your walkabouts.

However, do not discount other tools such as worker surveys and toolbox talks. The workforce may have strong or good ideas but may be too shy to approach you during a normal working day. Workers may prefer to come forward with their ideas during discussions with peers or as part of a survey, especially an anonymous survey.

3.3 Building relationships with the workforce

Active listening

Active listening can be one of the most important tools that a health and safety leader can have. Active listening will enable you to:
- successfully gather information. Leaders should be able to absorb, understand, and consider ideas and points of view from other people without interruption or argument; and
- listen to criticism without reacting defensively.

Good health and safety leaders need to learn to listen first and speak second so that meaningful, balanced communication always takes place.

Active listening is not the same as hearing. Hearing is about being aware of sounds. Listening requires action, you need to concentrate so that you process and understand the meaning of the message and can respond accordingly.

This is often represented using four stages: hearing, attending, understanding and remembering.

Hearing is simply being aware of a sound.

Attending is the act of filtering and screening, so you actually pay attention to the message. This is particularly important (and difficult) if you have issues with the person delivering the message. We need to listen through our prejudices and focus on the message not the messenger.

Understanding means we comprehend what is being said and decode the information we are being given. You should not just switch off until it is time for you to speak again. Listening properly is the only way we can really understand what people need, or what they think about a particular issue or idea.

Remembering means committing information to memory, an essential requirement for active listening. It is also the only way to ensure continuity and the important act of building rapport and a relationship. People like to be remembered and what they say is often important. It is not always easy to do, so make notes if needed. There is little point in a conversation, the sending and receiving of information through a communication channel, if the content of this exchange is instantly lost.

It is worth noting that the structure of your MBWA activity should involve both giving and receiving information if it is to have value.

Element 3 Leadership

3.3 Building relationships with the workforce

Encouraging improvement, innovation and learning (health and safety leadership value 5)

A 'learning organisation' is an organisation that listens, this means that any health and safety leader must also listen if improvement and innovation are to become truly embedded in the culture. Improvement and innovation can come from anywhere, the only way an organisation can learn about it is by ensuring everybody has a voice and that everybody is listened to. Feedback is as important as listening; workers are less likely to want to communicate if they get little or no feedback. An effective leader will know this and be creative in finding ways to develop a continuous process of encouraging innovation and improvement throughout all areas of the organisation. 'Managing by walking about' is one powerful method of gathering information across all levels.

It is important to ensure that development opportunities exist for all those who seek to learn. Professional development has proven to be a powerful motivation tool within many organisations. It also provides a platform for ideas for improvement and innovation to be proposed.

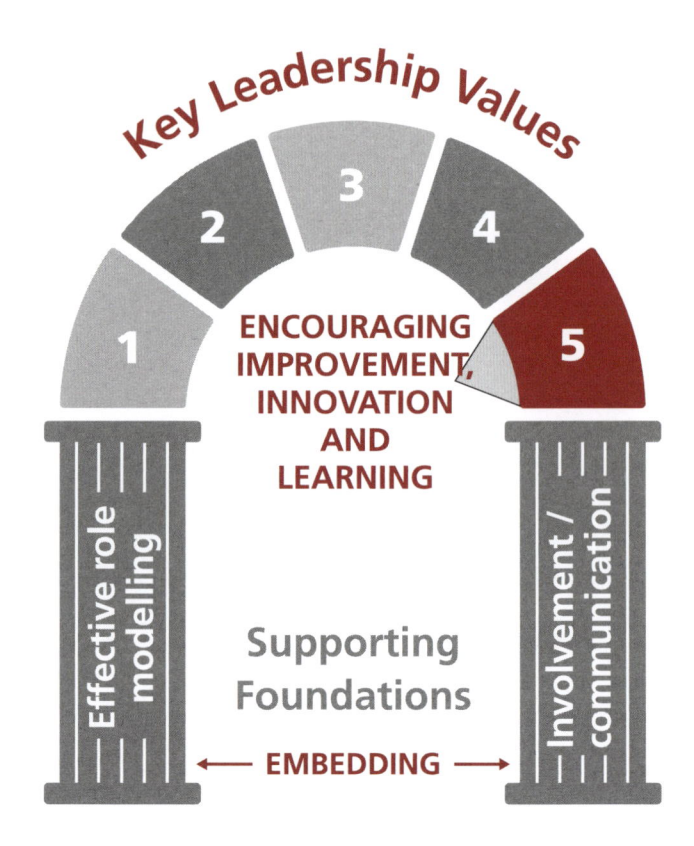

Assessment Activity

- Please refer to the document Unit HSL1, guidance and information for candidates and internal assessors.

You should now complete task L5: Encouraging improvement, innovation and learning.

Element 3 Leadership

3.3 Building relationships with the workforce

Positive reinforcement, negative reinforcement and punishment

> **Key Term**
>
> **Reinforcement**
>
> Reinforcement occurs when a consequence that follows a behaviour makes it more likely that the behaviour will occur again in the future.

Negative and positive reinforcement

The consequences of a behaviour is a driver for whether that action will happen again.

Negative reinforcement is avoiding a negative outcome or avoiding something that you do not want.

For example, if a worker knows that they will avoid a reprimand because their manager will walk past and not challenge them if they are not following procedures, then the act of walking past will negatively reinforce that unsafe behaviour. They are avoiding a reprimand and are likely to engage in the same behaviour again.

This is why the culture of the organisation is so important; the wrong culture can negatively reinforce the wrong behaviours.

Both positive and negative reinforcement will increase the chance that a behaviour will happen again.

Positive reinforcement is a technique that a good health and safety leader should be adopting to its fullest extent. It is important to reinforce positive behaviours (in this context, safe acts) so that they will happen again in the future. The best way to achieve this result is by praise. Everybody likes to receive praise, it makes us feel good; a simple 'thank you' can have a massive effect on a worker's day. So, for example, if you see a worker referring to a safe system of work before starting up a machine, you should highlight this by making a point of thanking the worker for following the procedures. This will not only make the worker feel good but will have an impact on their future behaviours. The worker will recognise the praise and will want to receive this in the future. This will lead the worker to repeat this action in future, even when the leader is not around. Not only will the positive reinforcement have a direct effect on the worker in question but it may also have an effect on other workers who may be in the area. They will have seen the worker receiving praise so will want the same for themselves. The result is that a simple 'thank you' could drive an improvement in safe behaviours throughout the workforce.

It is important for health and safety leaders to acknowledge that their workers will respond to a 'reinforcer'. This can be anything that is added to encourage the behaviour in the future. As well as praise and saying thank you, other reinforcers could include peer approval or disapproval and recognition in organisational newsletters. Effective reinforcers for a workforce can be identified through observation and by trialling potential reinforcers. For example, if everyone wears the correct defect-free personal protective equipment for the week, the workforce will be allowed to leave an hour early at the end of the week.

3.3 Building relationships with the workforce

It is also important to recognise that a reinforcer could be good for one person and bad for another. For example, a manager, who is a vintage car enthusiast, wants to thank his team for their safe working behaviours. The manager, therefore, arranges for the team to visit a vintage car show. The manager and car enthusiasts in the team have a really good time and will find this a good motivator for the future. However, the rest of the team have no interest in this subject and, therefore, feel the day was a tedious waste of time. This will not motivate the second half of the team to go that extra mile in the future.

We all too often only pick up on something when it goes wrong. However, the opposite is also very important; acknowledgment of safe behaviours should be at the front of health and safety leader's mind, especially when they are undertaking walkabouts or interacting with the workforce in some other way.

Punishment

> **Key Term**
> **Punishment**
>
> Involves an individual or group of workers receiving something they do not want (such as a verbal warning from a supervisor) and/or losing something that they have or would like to have (such as removal of bonus payments).

Punishment should only be used as a last resort but is appropriate for some situations. Punishment should always be consistently applied for the offence committed and be the same for all workers no matter what level they are in the organisation. For example, if a finance director refuses to wear personal protective equipment, they should receive the same punishment as a machine operator who has committed the same offence.

We all know that errors will always happen as we are only human. It is, therefore, important that the punishment is proportionate to the offence committed. It is essential that a full root cause analysis is carried out. This will determine the reasons for any errors or violations and the need to understand the local context of the individual at the sharp end of the failure.

Often errors and violations can be traced back to organisational failings such as, management decisions, time pressures, resourcing issues, lack of correct equipment and lack of suitable training. Individual workers should not be punished for errors or violations that are attributable to organisational failings.

Health and safety leaders should be aware that there are a range of punishments available. In reality, it is unlikely that the health and safety leader will implement the punishment, but they may be consulted on the type of punishment that is proposed to be given by the organisation's HR department. It is important to bear in mind the longer time effects any punishment could have on the worker. The worker is very likely already feeling guilty about the incident, especially if the incident has caused pain or longer term injuries to a colleague. The health and safety leader should be using judgement of what they know about the worker in question and the activity being undertaken. For example, if you know that a worker forgot to follow part of a procedure for machinery start-up due to arriving late at work after a sleepless night caused by a newborn baby, but has always diligently followed the rules in the past, it may be appropriate to advocate for a lesser punishment or even no punishment at all. However, if the worker is always ignoring procedures to cut corners to save time, you should be advising human resources of this so that an appropriate punishment can be given.

3.3 Building relationships with the workforce

In addition to internal actions being taken against a worker, there could also be legal ramifications. An individual, including a director, or a worker at any level, can be prosecuted in the UK for offences that breach health and safety legislation.

The following table summarises the main features of positive and negative reinforcement and punishment.

Positive reinforcement	Negative reinforcement	Punishment
Praise from health and safety leader	Avoid peer disapproval	Disciplinary action
Recognition from line manager	Avoid penalties	Removal of rewards
Approval of peers	Actions carried out are to avoid adverse consequences only	Suspension or dismissal

Element 3.3 References

44 Definition sourced from Oxford Languages.
45 HSE, 'Safe Deal Playing Cards', (https://books.hse.gov.uk/product/9780717666782/Pocket-Cards/Safe-Deal-Playing-Cards-pack-of-cards-Pack-of-cards)

Further reading

HSE, *Involving your workforce in health and safety. Guidance for all workplaces* (HSG263, 2014) (www.hse.gov.uk/pubns/books/hsg263.htm)

Sentencing Council, 'Manslaughter: Definitive guideline' (https://www.sentencingcouncil.org.uk/publications/item/manslaughter-definitive-guideline/)

Handy, C, *Understanding Organizations* (4th edition, 1993, Oxford University Press)

SHP Online, 'To err is human: human error and workplace safety' (2015) (https://www.shponline.co.uk/common-workplace-hazards/to-err-is-human-human-error-and-workplace-safety/)

Goleman, D, Boyatzis, RE, McKee, A, *Primal Leadership: Unleashing the Power of Emotional Intelligence* (2013, Harvard Business Review Press)

Health and Safety Authority, *Behaviour Based Safety Guide* (2013) (https://www.hsa.ie/eng/publications_and_forms/publications/safety_and_health_management/behaviour_based_safety_guide.html)